The Reconstruction of Post-War Labour Markets in The Southern African Development Community

Saint José Inaka ·
Christopher Changwe Nshimbi ·
Leon Mwamba Tshimpaka

The Reconstruction of Post-War Labour Markets in The Southern African Development Community

Insights from The Democratic Republic of The Congo

palgrave
macmillan

Saint José Inaka
Wits Institute for Social
and Economic Research (Wiser)
University of the Witwatersrand
Johannesburg, South Africa

Leon Mwamba Tshimpaka
SARChI Chair: The Political
Economy of Migration in the SADC
Region, Centre for the Study
of Governance Innovation (GovInn)
University of Pretoria
Pretoria, South Africa

Christopher Changwe Nshimbi
SARChI Chair: The Political
Economy of Migration in the SADC
Region, Centre for the Study
of Governance Innovation (GovInn)
University of Pretoria
Pretoria, South Africa

ISBN 978-3-031-37356-5 ISBN 978-3-031-37357-2 (eBook)
https://doi.org/10.1007/978-3-031-37357-2

Cover credit: ilbusca

This Palgrave Macmillan imprint is published by the registered company Springer Nature Switzerland AG
The registered company address is: Gewerbestrasse 11, 6330 Cham, Switzerland

To all workers in precarious working conditions, job seekers and survivors of post-war labour market misconstructions *in the Southern African Development Community (SADC) region and beyond.*

Foreword

The Covid-19 pandemic, starting in December 2019, had serious impacts on labour markets around the world. Many of its impacts continue, well after the disease has been contained. The impacts have generated various academic and policy debates and initiatives. Some of the debates consider how to address the problems and take advantage of the opportunities produced by the pandemic situation in the labour sector.

Apart from diseases like Covid-19, war and conflict affect labour markets too. The 10th edition of the *ILO Monitor on the World of Work* published by the International Labour Organization (ILO) on 31 October 2022 projects that global unemployment and inequality are going to rise because of various political and economic crises that threaten the recovery of the global labour market. According to the Monitor, declining labour markets are affecting employment creation and this indicates that the labour market is going to experience a sharp slowdown ahead.

Examining the world of work is important. However, very few studies consider the various impacts of events such as war on labour markets. This makes this book, *The Reconstruction of Post-war Labour Markets in the Southern African Development Community: Insights from the Democratic Republic of the Congo*, by Saint José Inaka, Christopher Changwe Nshimbi and Leon Mwamba Tshimpaka, uniquely significant.

The book makes a unique contribution to the understanding of the impacts of war on labour markets and efforts to reconstruct the labour markets in post-war periods. It focuses on neglected and understudied geographical and academic areas and epochs. It is the first work of its kind that examines the relationship between war and labour markets and is particularly focused on Southern Africa and the Democratic Republic of the Congo (DRC). Beyond the impact of war on labour markets, the book examines efforts to reconstruct labour markets in a post-war environment that yet again faces the threat of the recurrence of war. The historical stretch of the book, going back to just before the colonial era, adds to its unique flair and contribution to the academe. Southern Africa has been a theatre of various kinds of war throughout that period. Most significantly, the liberation wars fought in the region and lasting up to 1994, when South Africa got liberated from apartheid, means that the region can get along with the work of reconstructing its labour markets. This is despite the wars and conflict in some member countries of the Southern African Development Community (SADC) such as the DRC.

Among the various episodes of war it has experienced, the DRC experienced a severe one from 1998 to 2002. That war seriously affected the labour market of the country besides other spheres of the Congolese society and economy. Its effects on the country's labour market have generated various debates. *The Reconstruction of Post-war Labour Markets in the Southern African Development Community: Insights from the Democratic Republic of the Congo* offers a first analysis of the process of reconstructing the labour market in post-war DRC, within the context of the SADC region, and of the conscious or oblivious roles played by selected key actors in the reconstruction.

In responding to the fundamental question about the processes involved in the reconstruction of the post-war labour market in the DRC, and how selected key actors affected those processes, the book makes significant theoretical and empirical contributions to the understanding of labour markets. The contributions show that the post-war labour markets in Southern Africa, and the DRC in particular, have undergone extended reconstruction. The reasons for this are detailed in this original study, which uses and demonstrates the utility of theories such as Peck's theory of labour market social regulation in locally specific ways as theoretical tools for evaluating problematic transitions from war to viable labour markets. Indeed, the book provides fresh insights into how both state and non-state institutions as well as state and non-state actors work to rebuild labour markets destroyed by war.

March 2023

Prof. Dr. Patrick Develtere
Professor of International
Development Cooperation
University of Leuven (KU
Leuven)
Leuven, Belgium

Acknowledgements

This book and the research on which it is based are the result of the assistance of many people. Among them were the various respondents and discussants from different backgrounds, who fully supported the work by making themselves available for interviews and focus group discussions. The book is also a result of fruitful collaboration and the support of a number of institutions and entities within those institutions. We gratefully acknowledge the support of the National Research Foundation (NRF), the University of Pretoria (UP) and the University of the Witwatersrand (Wits). At UP, we particularly acknowledge the Centre for the Advancement of Scholarship (CAS), the Human Economy Research Programme therein as well as the Department of Sociology. We also acknowledge the Department of Political Sciences, the Centre for the Study of Governance Innovation (GovInn) and the SARChI (Southern African Research Chairs Initiative Chair) Research Chair in the Political Economy of Migration in the SADC region in that department. We wish to express our sincere gratitude to Professor James Ogude, Director of CAS, Professor John Sharp, Director of the Human Economy Research Programme and Professor Victoria Graham, Head of the Department of Political Sciences. We are also grateful to Professor Maxi Schoeman, Professor Jonathan Hyslop, Professor Debby Bonnin, Professor Zitha Mokomane, Professor Keith Hart, Professor Fraser Mc Neil, Dr. Joe Trapido, Dr. Tafadzwa Mushonga, Dr. Sepetle Molapo, Dr. Detlev Krige, Dr. Sandla Nomvete, Dr. John Mashayamombe, Dr. Wallace Numvi, Dr. Jose Adalima and

Dr. Amanuel Towelde, all at UP, for inspirational leadership, collegiality, advice and support. At Wits, we are grateful to the Wits Institute for Social and Economic Research (WISER) and particularly to Professor Achille Mbembe. Our gratitude also goes to Professor Leon Tsambu, Professor Jean-Marie Mutamba and Professor Baby Ilunga of the University of Kinshasa; Professor Pedro Monaville and Professor Joshua Walker at the University of New York and Professor Nancy Hunt of the University of Florida. Your valuable advice is deeply appreciated. We also wish to thank the various anonymous reviewers who gave us valuable suggestions and critically reflected on the manuscript in the multiple rounds of review it went through. Finally, our special thanks to friends and family and, indeed, our respective beloved spouses, Mrs. Fanny Inaka, Mrs. Charity Nshimbi and Mrs. Jolie Tshimpaka. Our children are not forgotten and especially Dr. Inaka's daughter, Santa Maria Laetitia Inaka.

Pretoria and Leuven
March 2023

Saint José Inaka
Christopher Changwe Nshimbi
Leon Mwamba Tshimpaka

Contents

Abbreviations and Acronyms

AACP	Anglo-American Council on Productivity
ABIR	Anglo-Belgian India Rubber and Exploring
ACP	*Agence Congolaise de Presse* (Congolese Governmental Press Agency)
AFDL	*Alliance des Forces Démocratiques pour la Libération du Congo* (Alliance of Democratic Forces for the Liberation of Congo)
AFL	American Federation of Labor
ANAPI	*Agence Nationale pour la Promotion des Investissements* (National Agency for the Promotion of Investments)
ANC	*Armée Nationale Congolaise* (Congolese National Army)
ANEZA	*Association Nationale des Entreprises du Zaïre* (National Association of Zairian Enterprises)
ANR	*Agence Nationale des Renseignements* (National Intelligence Agency)
APIC	*Association de Personnel Indigène de la Colonie* (Association of Indigenous Personnel of the Colony)
APIPO	*Association des Postiers de la Province Orientale* (Oriental Province's Native Postmen's Association)
ATC	*Alliance des Travailleurs du Congo* (Alliance of the Congolese Workers)
AU	African Union
BIAC	*Banque International de l'Afrique Centrale* (International Bank of Central Africa)
BLS	Bureau of Labor Statistics, United States Department of Labor
CBT	Belgian Technical Cooperation

CDT	*Confédération Démocratique du Travail* (Democratic Confederation of Labour)
CGSA	*Confédération Générale des Syndicats Autonome* (General Confederation of Autonomous Unions)
CIAT	*Comité International d'Accompagnement de la Transition* (International Committee in Support of the Transition)
CIPRAP	*Commission Interministérielle de Pilotage de la Réforme de l'Administration Publique* (Interdepartmental Steering Committee for the Reforms of the Public Administration)
CNSS	*Caisse Nationale de Sécurité Sociale* (National Social Security Fund)
CNT	*Conseil National du Travail* (National Council of Labour)
COMESA	Common Market for Eastern and Southern Africa
CSG	*Confédération Syndicale du Congo* (Congolese Trade Unions Confederation)
CSGA	*Confédération Générale des Syndicats autonomes du Congo* (Congolese General Confederation of Autonomous Trade Unions)
CTB	*Coopération Technique Belge* (Belgian Technical Cooperation)
DG	Director General
DGI	*Direction Générale des Impôts* (Directorate-General for Taxation)
DGRK	*Direction Générale des Recettes de Kinshasa* (General Directorate of Revenue Collections of Kinshasa)
DRC	Democratic Republic of Congo
ECA	Economic Cooperation Administration, United States of America
ECA	United Nations Economic Commission for Africa
ECM	Extended Case Method
EDUCAT	*Éducation au Katanga* (Education in Katanga)
ERPTUAC	European Recovery Programme Trade Union Advisor Committee
FAO	Food and Agriculture Organisation, United Nations
FARDC	*Forces Armées de la République Démocratique du Congo* (Armed Forces of the DR Congo)
FEC	*Fédération des Entreprises du Congo* (Congolese Business Federation)
FGTK	*Fédération Générale du Travail du Kongo* (General Federation of Labour of Kongo)
FLNC	*Front pour la Libération Nationale du Congo* (Front for the National Liberation of Congo)
FNLA	National Front for the Liberation of Angola
FOSYCO	*Force Syndicale du Congo* (Congolese Union Force)
FRELIMO	Mozambique Liberation Front
GIZ	German Society for International Cooperation

IBTP	*Institut des Bâtiments et des Travaux Publics* (Building and Public Works Institute)
ILO	International Labour Organization
IMF	International Monetary Fund
INPP	*Institut National Préparation Professionnelle* (National Institute of Vocational Training)
IOM	International Organization for Migration
IRCA	International Railways of Central America
JICA	Japan International Cooperation Agency
MLC	*Mouvement pour la Libération du Congo* (Movement for the Liberation of Congo)
MNCs	Multinational Corporations
MONUC	*Mission des Nations-Unies au Congo* (Mission of the United Nations in Congo)
MONUSCO	*Mission de l'Organisation des Nations Unies pour la Stabilisation en République Démocratique du Congo* (United Nations Organization Stabilization Mission in the DR Congo)
MPLA	Popular Movement for the Liberation of Angola
MPR	*Mouvement Populaire de la Révolution* (Popular Movement of the Revolution)
NGO	Non-Governmental Organisation
OEEC	Organisation for European Economic Cooperation
ONEM	*Office National de l'Emploi* (National Employment Office)
ONUC	*Opération des Nations Unies au Congo* (United Nations Operation in the Congo)
OPC	Ovamboland People's Congress
OPDSC	Organ on Politics, Defence and Security Cooperation
OTUC	*Organisation des Travailleurs Unis du Congo* (Congolese Organisation of United Workers)
PPRD	*Parti du Peuple pour la Reconstruction et la Démocratie* (People's Party for Reconstruction and Democracy)
RCD	*Rassemblement Congolais pour la Démocratie* (Congolese Rally for Democracy)
SADC	Southern African Development Community
SAP	Structural Adjustment Programme
SENEM	*Service National de l'Emploi* (National Employment Service)
SIAP	*Syndicats Indépendants de l'Administration Publique* (Independent Unions of the Public Administration)
SNEL	*Société Nationale d'Électricité* (National Electric Company)
SOPA	*Solidarité Ouvrière et Paysanne* (Worker and Peasant Solidarity)
SWAPO	Southwest African People's Organisation
TOE	*Les travaux d'ordre educatif*
UFCO	United Fruit Company

UMHK	*Union Minière du Haut-Katanga* (Mining Union of the Upper-Katanga)
UNDP	United Nations Development Programme
UNESCO	United Nations Educational, Scientific and Cultural Organisation
UNHCR	United Nations High Commissioner for Refugees
UNICEF	United Nations Children's Fund
UNITA	National Union for the Total Independence of Angola
UNTC	*Union Nationale des Travailleurs du Congo* (National Union of the Workers of the Congo)
UNTZA	*Union Nationale des Travailleurs du Zaïre* (Zairian National Union of Workers)
USA	United States of America
USAID	United States Agency for International Development
WWI	World War I
WWII	World War II

List of Figures

PART I

Historical and Theoretical Bases of Post-War Labour Market Reconstruction in the SADC Region

CHAPTER 1

On Post-war Labour Market Reconstructions in Southern Africa

Southern Africa engaged in wars of liberation from colonial rule in the late to mid-twentieth century. The region fought the wars to gain independence and the right to self-determination. Upon attaining political independence, some countries in the region descended into civil war and conflict, as various factions within their territories contended for power or the new rulers sought to consolidate state power. As of mid-2021, however, no active armed conflict existed within each of the member states of the Southern African Development Community (SADC). The only exceptions were the Democratic Republic of the Congo (DRC) and Mozambique.

Having obtained political independence, all SADC countries aim to achieve sustainable development in line with the regional goals of integration. They aim to do so by, among other things, promoting good governance and democracy, peace and security, the management of conflict, post-conflict reconciliation, reconstruction and rehabilitation. While most of the literature that examines these efforts tends to focus on trade and democracy, very little of it deals with labour markets. Even the research that focuses on African labour markets itself pays little attention to post-war labour market reconstruction. This is despite the fact that almost all African countries have experienced war or some form of conflict in the past. This chapter introduces the book by highlighting this

S. J. Inaka et al., *The Reconstruction of Post-War Labour Markets in The Southern African Development Community*,
https://doi.org/10.1007/978-3-031-37357-2_1

gap in the literature on (the history of) post-war labour market reconstruction and labour market recovery in post-war countries around the world and in Southern Africa. In so doing, the chapter also reveals the novel empirical and theoretical contribution of the book to studies on labour market recoveries and labour market theories in disciplines such as Sociology, Political Science and Economics. The question could be asked, whether the attempt to examine and use the notions 'reconstruction' and 'recovery' in an undertaking such as this one is not too optimistic. The question arises from doubts about whether a viable labour market exists in the DRC or whether the country has any labour market at all. To this the point may be made that, while the Government of the DRC is argued to be the largest formal employer in the country, the fact that multinational corporations (MNCs) operate in the DRC suggests that and points to a viable labour market, whether formal or informal. Moreover, in the context of regional organisations such as SADC, the Common Market for Eastern and Southern Africa (COMESA), the African Union (AU) as well as international non-governmental organisations (INGOs) such as the International Labour Organization (ILO), United Nations Economic Commission for Africa (ECA), United Nations Development Programme (UNDP), United States Agency for International Development (USAID) and the International Organization for Migration (IOM), Africa is poised to grow a labour market that seeks to meet the aspirations of its burgeoning youthful population. All of this speaks to 'reconstruction' and 'recovery', especially in view of and following the devastations of (recurring) wars in countries like the DRC. Hence, this proposed book. This chapter also outlines the theoretical perspectives, methods employed and the contents of the book.

The rest of the chapter is organised in five sections. The first provides a contextual diachronic overview of post-war labour market reconstruction around the world. The second zeroes in on to the SADC region, to demonstrate the dialectical relationship between war and labour market problems. The section particularly focuses on labour market developments that caused and fuelled the wars in the region. Section "The Origins of SADC and the Apathy Towards Post-war Labour Market reconstruction in Southern Africa" explains why the SADC region is doing so little to address issues of post-war labour market reconstruction, when it and the respective countries are clearly impacted. Section "Methodological and Theoretical Considerations" introduces the objectives and discusses the methodology deployed and theoretical approach of the book. Section "The Structure of This Book" outlines the structure of the book.

Post-war Labour Market Reconstructions: A Diachronic Global View

Although the 1919 Treaty of Versailles acknowledges the importance of jobs after wars (Data-Bah 2003: 3), many studies on war-torn countries and countries that are emerging from war tend to focus on economic recovery and political issues like democratisation and not on post-war labour markets (Cramer 2008). A neglected aspect in the study of labour from a disciplinary perspective and particularly from a sociological position is labour in situations of war, violence and immediately after war (Giddens 1985; Brewer 2010, 2013; Malešević 2010b; Joas and Knöbl 2012). A reason for this neglect is the assumption by Sociology's founding thinkers that war and violence are irrational and primitive, and doomed to vanish as society progresses to modernity (Giddens 1985; Tilly 1985; Mann 1988). Similarly, American and European scholars of the 1960s and 1970s avoided studies of war because of the widespread anti-war sentiments after the experiences of World War I and World War II, which were manifested in relation to the Vietnam war (Malešević 2010; McSorley 2014).

Unlike most sociologists, Johan Galtung, the pioneer of peace studies and the world's most experienced peace scholar (Brewer 2013; McSorley 2014), has taken a keen interest in these studies. Historians, political scientists and specialists of military studies, rather than sociologist, follow and build on Galtung's work that theoretically distinguishes negative peace from positive peace[1] (Galtung 1969), analyses various challenges linked to peace processes (Galtung 2004; Webel and Galtung 2007), the rise of international terrorism (Galtung 2005) and structural and cultural violence (Galtung 1990). These researchers, who represent diverse fields, have conducted relevant studies on post-war market reconstructions. These studies focused on the Cold War era and demonstrated how both capitalist and communist ideologies influenced these processes. They outline the actions of the USA and the Soviet Union, and the state's role

[1] In Galtung's understanding, negative peace only means the absence of violence; positive peace signifies the achievement of fairness, justice, democracy, development and social redistribution.

in post-war countries and how they influenced changes in labour market regimes during the Cold War.

Labour Market Reconstructions Post-World War II and the Cold War Era

As a superpower, the USA was involved in the post-war reconstruction of its ally and client countries (Del Castillo 2008; Sorel et al. 2008) in Europe. It helped European countries to recover economically through the European Recovery Program (ERP), also known as the Marshall Plan (Toropov 2000; Wasser and Dolfman 2005; Machado 2007; Del Castillo 2008). This resulted in changes in labour market regimes of European countries due to the implementation of labour reforms and regulations based on US Fordist labour regulations and Keynesian principles (McKenzie 2007).

In addition to its direct involvement in Europe, the USA also supported democratic regimes like Japan (Yokoi 2004), many totalitarian regimes in Asia and Latin America (Nye and Owens 1996; Levitsky and Way 2010) and kleptocratic regimes in Africa such as Zaïre under Mobutu and Liberia under Samuel Doe (Englebert and Tull 2008). US funding of Japanese reconstruction enabled the Japanese government to introduce a 'Japanese-style management' labour regime (Forsberg 2000; Yokoi 2004) based on consensual collaboration between government, employers, trade unions and workers (Forsberg 2000). Although the USA also funded the militaristic regime in South Korea after the Korean War of the 1950s (Webster et al. 2008), the South Korean government collaborated with the *chaebols* (family-owned conglomerates) to set up an authoritarian militaristic labour market regime. The regime was characterised by anti-trade unionism policies (Kwon and O'Donnell 2003) and militarised industrial relations (Webster; Lambert and Beziudenhout 2008). In Africa, post-war nation-states subjected trade unions and employees to domination by political parties and the state (Kester 2007; Britwum 2012). Overall, post-war labour reconstructions in capitalist countries were significantly influenced by the USA out of concern with the communist threat to its political and economic interests. This made the USA intervene in the countries where it did so, on humanitarian, economic and/or geopolitical grounds. Recovery of the labour market in those countries during the Cold War occurred through the interaction of many different factors and actors, on the basis of a capitalist market.

The socialist bloc, however, organised labour market reconstruction after the war in a different way. Generally, labour market reconstruction in the Soviet Bloc and allied countries was made dependent upon governments. The governments portrayed themselves as the 'sole owners' of all political, ideological and economic powers, and as the controllers of the labour market. Communist policies thus shaped the labour market reconstructions and regulations. Yet global geopolitics negatively affected the functioning of those labour markets. For one, the economy of the Soviet Union was severely affected by World War II. And while it was reconstructing its own economy, it also assisted its allies in the Soviet Bloc (Harrison 2002) to reconstruct their own labour markets (Porket 1995). These endeavours were often disrupted by economic embargos imposed by the USA, as in Cuba and Vietnam for example. In Cuba, Fidel Castro ended the country's economic dependence on the USA by aligning with the Soviet Union (Alexander 2002). He nationalised the economy, including all privately held land, companies and private property (Magnusen and Rodríguez 1998). Fidel Castro strived to improve Cuban living standards by increasing wages (Chomsky 2011).

After their victory, the Vietcong in Vietnam nationalised all private companies (Joes 2001) and consolidated alliance with the Soviet Bloc. Vietnam developed a five-year economic plan that focused on developing industry and collective agriculture. It then implemented an anti-trade unionism labour regime and created New Economic Zones (Gustafsson 2010). In the meantime, public service workers, religious and military officers of the former government were imprisoned and forced into heavy and risky work—essentially slave labour—in re-education camps (Truong 2014).

The policies introduced by the Cuban and Vietnamese governments as well as the embargos imposed by the USA had serious repercussions for the economies and labour markets in these two countries. In Cuba, the government struggled to pay workers (Alexander 2002), as a result of which Castro introduced a labour policy that encouraged non-paid voluntary work (Chomsky 2011). Vietnam's experience was similar. During the mid-1980s, the country faced severe shortages of food and basic necessities, and hyperinflation (Vuong 2014; Tran 2017). In the late 1980s, a five-year-plan was developed (Vuong 2014), which led to the abandonment of collective industry and agriculture and the adoption of a market economy. With this, the Vietnamese labour market regime shifted from one of slave labour to one of unionised and cheap labour (Tran 2017).

Both the Soviet Union and the USA, respectively, supported their allies in wars and, later, in governing their war-torn countries, whether directly or indirectly. The outcome was post-war labour markets grounded in a market-oriented economy (for countries that adopted capitalist ideology) or grounded in a state-led economy (for those that followed communist ideology). The geopolitical clashes at the world level thus interfered with and shaped local political governance and the reconstruction of labour markets in various countries around the world.

Scholarly Literature on Post-Cold War Labour Market Reconstruction and Post-war Labour Market Recovery

The scholarly literature also examines the rebuilding of post-war countries in the post-Cold War era. Two themes determine labour market reconstruction during this period: *causa belli* and *modus vivendi*. According to Kalyvas and Balcells (2010) and Kaldor (2012), since the end of the Cold War in 1989, conflict no longer takes the form of war between states that results from political and/or ideological differences. It instead tends to be intra-state civil war. This shift relates to changes in the *causa belli*, though the exact cause is subject to debate.

One view in these debates maintains that civil wars that have occurred since the end of the Cold War have been driven by the economic agendas of warlords, criminals and thugs (Collier and Hoeffler 2004; Weinstein 2005; Brewer 2010; Kevlihan 2013). This view is based on Collier's (2000) theory of greed and grievance, by which the belligerents wage war, loot state resources, extort people and engage in national and international criminality for self-enrichment. A contrary view posits that greed theory does not fully capture the dynamics of post-Cold War conflicts (Stewart 2008; Ugarriza 2009; Keen 2012). Rather, grievances and issues concerning religious or ethnic discrimination cause war (Fearon and Laitin 2003). Stewart and Langer (2008) and Stewart (2016) speak of 'horizontal inequalities' or inequalities between ethnic, political, employment, religious or racial groups as factors that play an important role in perpetuating armed conflict in developing countries.

The *modus vivendi* of post-war countries has changed too, with the end of the Cold War (Kalyvas and Balcells 2010). The main change in this regard is that international partners, such as United Nations peacekeeping missions (Paris 1997), international development agencies, non-governmental organisations (NGOs), international financial institutions and regional organisations mimic the role of the Cold War superpowers (Kaldor 2012).

Furthermore, most international partners in post-war labour market reconstruction processes base their work on neoliberal ideologies (Izzi 2013). The implementation of neoliberal labour market policy reforms has resulted in burgeoning academic literature focused on the impact of these policies on labour markets. The key principle of neoliberalism is that the labour market is grounded in a market-oriented economy, defined by Roland Paris as:

> [a]n economic order in which goods and services are predominantly produced and allocated by more or less competitive firms that are predominantly privately owned and strongly influenced by market prices and by the goal of profitability. (Paris 1997: 56)

In the context of post-war labour market reconstruction, neoliberal reforms, mostly sponsored by the World Bank and the International Monetary Fund (IMF), aim to privatise public enterprises, reduce governments' intervention in the functioning of the labour market, improve the business climate and establish the rule of law for commercial relations and an efficient fiscal system (Cubitt 2011).

However, some empirical studies (Data-Bah 2003; Cramer 2008; Stewart 2008, 2016) demonstrate that these neoliberal policies have negative consequences for the post-war labour market. The loss of jobs (Stewart 2012), increase of precarious jobs, abuse of workers' rights and weakening of trade unions are all indicated as consequences of neoliberal policies (Izzi 2013).

The end of the Cold War saw the conclusion of many civil wars in Africa and other parts of the world, with warring factions entering peace negotiations. These peace processes were generally implemented in two ways. The first involved dividing war-torn countries such as Eritrea's secession from Ethiopia in 1993, East Timor's split from Indonesia in 2002 (Collier 2011) and Yugoslavia's break up into seven different countries (Beasley 2006; Stewart 2008, 2012). The second involved the creation of transitional institutions in order to promote power sharing within nation-states (Mehler 2009; Simons et al. 2013). The power-sharing provision is widely used as a transitional mechanism in post-war African countries (Levitt 2012), where former adversaries compromise to create a viable structure of shared rulership in their countries for a given period (Hartzell and Hoddie 2007). Power-sharing accords are political agreements between adversaries that formally outline the way in which power will be shared in the legislative and/or executive branches of government—including both formal or informal provisions for political, military, economic and territorial functions (Jarstad 2008; Binningsbø and Dupuy 2009). However, the conceptualisation and implementation of power-sharing processes in Africa often ignore domestic, regional and international law, doctrine, norms and jurisprudence (Levitt 2012). Although power sharing might strengthen peace and foster democracy (Jarstad 2008), it is also a source of political instability, ineffective governance and violent conflicts (Tull and Mehler 2005). The literature on post-war reconstruction engages a lot with such narratives. What one, however, observes as a gap in this literature on especially the *modus vivendi* of post-war countries is that much has been written on the role of political authorities but little of this research has been done on post-war labour market reconstruction in the context of power-sharing provisions. As power-sharing provisions govern both post-war countries and their labour markets, it is important to understand how such political regimes shape the remaking of labour markets. However, the literature does not show how such power-sharing provisions influence labour market reconstruction. This is problematic since labour markets need to be considered, first and foremost, as a political construct (Peck 1992). It is very likely that if power-sharing provisions shape peacebuilding processes and the governance of post-war countries, it would also impact labour market reconstructions. Therefore, it is of profound interest to focus on this gap.

Besides, it is also important to recognise that some scholars have examined policies for post-war labour market reconstruction and issues of labour supply, labour demand and labour market institutionalisation. In a post-war setting, labour markets can become chaotic due to an absence of labour legislation and regulation, which in turn can cause inequitable employment practices (Data-Bah 2003). Reconstructing such labour markets can be challenging, in particular for developing countries (Beasley 2006). The reasons for this are twofold. First, the economies of most such countries are weak (Bray 2005). Second, the absence or weakness of state institutions often gives rise to unemployment, underemployment (Bray 2005) and job insecurity (Cramer 2008). Hence, it is necessary to focus on labour market policy reforms, labour supply/demand reform programmes and labour market institutional issues when studying post-war labour market reconstruction (Data-Bah 2003; Beasley 2006; Cramer 2008).

It has been argued that labour market policy reforms are crucial for peace and job creation in order to improve socioeconomic conditions, dignity, social healing and hope in post-conflict countries (Izzi 2013). For that reason, state authorities in war-torn countries often work together with international partners, particularly the World Bank and the IMF, to conceive and find funding for such reforms (Data-Bah 2003; Cramer 2008, 2011; Stewart 2012, 2016). Where these reforms have been successful and what they have achieved is, however, debatable. Four points of contestation are particularly worth highlighting.

First, claims of success have been critiqued for assuming that post-war employment policies inherently lead to improvements in the labour market (McLeod and Davalos 2008; Amarasuriya et al. 2009). For example, state authorities in Bosnia and Herzegovina together with the World Bank celebrated an employment programme for reducing the unemployment rate from 64 to 44% (Stewart 2016). However, empirical studies show that the programme did not lead to improvements in the labour market: employment rates remained low because of the persistence of restricted access to well-paid permanent jobs, poor labour protection of employees and labour inequalities based on ethnicity and gender (Beasley 2006; McLeod and Davalos 2008). The situation in Kosovo looked quite similar. A 2008 World Bank report posits that the Kosovo Youth Employment Action lowered the unemployment rate from 44.6 to 35%. However, an empirical study by Stewart (2015: 12) found that the programme did not include Serbs due to their ethnicity and that it created few job opportunities for skilled people. And in Sri Lanka, local state authorities, supported by the United Nations (UN), set up an Action Plan for Youth Employment in 2000 and then presented this as having led to a reduction in the unemployment rate from over 60 to 40%. However, empirical studies yet again showed that 20% of those categorised as 'employees' were in fact underemployed (Amarasuriya et al. 2009). Based on these examples, it is plausible to argue that what matters more than an increase in the number of jobs is the nature of employment, its coordination and the resulting effectiveness of labour market actors, and overall efficient labour market regulation.

Second, claims to success have been queried by pointing to the fact that economic growth is inversely linked to an increase in unemployment. In Sierra Leone (Cubitt 2011) and Mozambique (Paris 2004), for example, the governments aimed to stimulate economic growth by reducing their roles in the economy, improving the business climate and privatising public enterprises, following the recommendations of the IMF and the World Bank. Yet, while both countries achieved economic growth (Paris 2004), this was not accompanied by the creation of new jobs in the formal sector while labour market inequalities, exploitation, underemployment and poverty remained unchanged (Cubitt 2011). In this regard, the literature clearly demonstrates the limitations of neoliberal assumptions that liberalisation, privatisation and deregulation of post-war labour markets create job opportunities.

Third, the literature identifies failures in both how the local authorities and their international partners deal with reforms. For example, local authorities obstructed the implementation of reforms in the DRC (Diumasumbu 2008; Trefon 2010, 2011), Afghanistan (Barakat 2002) and Sierra Leone (Cubitt 2011) and managed them poorly in Mozambique (Dunne 2003) and Uganda (Stewart 2015). International partners, in turn, fail to conduct in-depth investigations into the specific political, economic and sociocultural factors that shaped labour markets on the ground (Dunne 2003).

Fourth, international agents often do not share the same views (Dunne 2003). It is surprising to note that there is often a lack of coordination, cohesion and understanding between international partners, such as the ILO, the World Bank and the IMF. These international organisations are supposedly joint authors of reconstructed labour markets. Yet, they are often consumed by internecine wars. They often hold opposing views because objectives concerning employment differ: where the ILO advocates an increase and protection of employment, based on its Decent Work Agenda, the World Bank and the IMF tend to focus on promoting high employment through sustainable economic growth. Often, the decision of the kind of advice a country will implement is determined by the capacity of the World Bank and the IMF to lend money (Berg and Kucera 2008).

Beyond these contestations, supply policies generally focus on improving systems of information for employment by setting up employment services and building the capacities of the workforce through vocational training. Cubitt (2011) shows that those in Sierra Leone did not assist low- or semi-skilled workers. The second way to build labour supply is to increase employability. This is by instituting vocational training, internships and apprenticeships (Date-Bah 2006; Stewart

2015). The reason for this focus is that short courses allow people to earn a living more quickly while higher education is expensive and time-consuming. Often vocational training is organised through Disarmament, Demobilisation and Reintegration (DDR) programmes and employment promotion programmes for women and youth. Many positive impacts have been observed for the DDR programmes in Uganda (Inaka 2020), Sierra Leone (Cubitt 2011), Mozambique (Cramer 2008) and Cambodia (McLeod and Davalos 2008), though Krishnamurty (2007) points out that their assistance of ex-combatants neglects other unemployed people who are also affected by war.

The promotion of women's employment through vocational training programmes has also proven to be problematic. Vocational training was introduced in Mozambique, Guatemala, Lebanon and Bosnia to develop women micro-entrepreneurs (Data-Bah 2003). The programmes were sponsored by international partners (like the United Nations High Commissioner for Refugees (UNHCR), the United Nations Children's Emergency Fund (UNICEF) and the USA) and local non-governmental organisations (NGO). Yet the training focused on a very small selection of skills, namely sewing, hairdressing, cake decoration and knitting. As a result, the women who underwent the training found themselves in direct competition with each other, producing similar goods, which lowered their ability to earn proper incomes from their work—while nevertheless bearing the traditional stigma of being working women.

A similar pattern was found in studies of youth employment programmes in Sierra Leone (Cubitt 2011) and Guinea Bissau (Izzi 2013). Although young people received practical training in building, mechanical maintenance, electricity, driving and running microenterprise projects, few of them were subsequently able to find employment. The vocational training offered was driven mainly by the preferences of international funding partners. But the programme contents were often disconnected from the political and economic realities on the ground, not to mention the actual needs of the local labour market.

There are also labour demand policies on the other side. These policies focus on creating jobs in the public sector and on promoting investment that, in turn, would create job opportunities in the private sector (Izzi 2013). In practice, however, several studies have found that labour demand policies tend to create very few job opportunities, whether in the public or the private sectors of post-war countries and that, indeed, possibilities of job creation are quite limited (Bray 2005; Cramer 2008; Izzi

2013). One of the problems behind this is that little attention is paid to what is actually required on the ground: they are not conceptualised to meet local economic, political and employment realities. Another is that many private investors are reluctant to invest and create jobs in post-war countries because of unsound business climates and security risks (Cubitt 2011). The only industries prepared to risk investment in post-war countries are mining, petroleum, mobile communication, construction and engineering—and some commercial banks (Bray 2005). In addition, some of the companies that do want to invest, want to do so exactly because the countries are disorganised after a war: these are not companies that necessarily want to act for the good of the labour market. That is why an increase of investment does not necessarily lead to job creation.

Overall, the existing but scarce literature on post-war labour market reconstructions, both after WWII and after the Cold War, has provided a brief historical global examination of the policies of labour demand, labour supply and labour market institutionalisation that tend to be applied in labour market reconstructions. Within this literature, it is demonstrated how capitalist and communist ideologies had a bearing on post-war labour market recoveries during the post-WII and Cold War eras. However, the ways in which neo-colonialism has influenced and influenced these processes have been insufficiently studied. The shift to the post-Cold War period led to accompanying changes in attempts to understand the causes of wars and how rebuilding processes were undertaken in post-war countries and their labour markets. This period was marked by reconstructions that were based on neoliberal ideologies and that strengthened asymmetrical relations between local and international actors. The following section attempts to show these realities through relationships between labour markets and wars in the SADC region.

Overall, the question that we are faced with is, if countries that have moved on the way past war times have often shown serious shortcomings in working through post-war labour market reconstruction processes, how much more (or less) could countries in the SADC region navigate such processes? This question necessitates an overview of relationships between labour market issues and wars of liberation from colonial rule and apartheid, the right to self-determination and racial (in)equality in Southern Africa. As a follow-up, it is also necessary to succinctly explain how labour market issues shaped civil wars in some post-independent SADC member countries such as Angola, Mozambique and the DRC. These questions are addressed in depth in the following sections of this chapter and in the rest of the book. Therefore, before outlining the structure of the book, the next three sections successively engage with the relationship between labour markets and wars that have occurred in the SADC region, including why despite this, the SADC region does not seem to engage in post-war labour market reconstruction.

Independence, Consolidation of State Power, Civil Wars/Conflict and Labour Market Reconstructions in the SADC Region

Almost all the countries of Southern Africa went to war to fight against colonial rule or apartheid post-WWII up to the early 1990s. These wars were shaped by both internal and external political, psychological, sociocultural and economic factors (Alexander et al. 2017: 2; Bvirindi 2019: 1). In terms of external factors, several foreign interferences ignited and fanned these wars, which manifestly epitomised ideological conflicts between capitalism (led by the USA) and communism (claimed by the Soviet Union) in Africa during the Cold War (Alexander et al. 2017: 2–4). Internally, these wars had their grounds in cultural cleavages and socioeconomic inequalities including, racism, ethnicity, gender discrimination and class conflicts (Greenberg 2008; Shubin 2008). Though all these factors contributed to wars fought between forces backed by the Soviet Union and the USA, it is crucial to demonstrate how labour market realities—such as racialised or gendered labour market discriminations, forced labour, underpayment, unemployment—were also among important drivers of these wars.

In the interest of space, we do not engage in a detailed discussion of all the different and complex connections between labour market realities and wars in the 16 diverse countries of Southern Africa that constitute the SADC. It is, however, crucial to underscore the dialectic relations between labour market realities and wars. In other words, it is important to demonstrate the ways in which labour market issues contributed to these wars, and how the wars disarticulated the labour markets that needed to be reconstructed in post-war times.

For starters, the SADC region includes the Lusophone countries of Angola and Mozambique, apart from the Anglophone and Francophone. Three liberation movements in Angola namely, the Popular Movement for the Liberation of Angola (MPLA) led by Augisthino Neto, the National

Front for the Liberation of Angola (FNLA) led by Holdern Roberto and the National Union for the Total Independence of Angola (UNITA) led by Jonas Savimbi fought the Portuguese colonisers for liberation from 1961 until Angola's independence in 1975 (Bvirindi 2019: 2). These three liberation movements were influenced by interconnected internal and external events in waging war (Leão and Rupiya 2005: 8). The emergence of several African liberation movements, the rise of Pan-Africanism, the advent of the international bipolar system and the wave of independence in Africa around the 1950s were among major factors that awakened Angolan political consciousness and nationalism (Guimaraes 1992; Leão and Rupiya 2005). On the domestic front, the effects of oppressive labour market policies and regulations (the excesses of forced labour, the exploitative contract labour system, racialised labour discriminations) were among factors that pushed many Angolans into engaging in armed struggle against the Portuguese colonists (Weigert 2011: 32–34).

In Mozambique, the liberation war against the Portuguese was led by the Mozambique Liberation Front (FRELIMO) from 1964 until the end of colonial rule in 1975. International events similar to those affecting Angola were behind this war including, the rise of anti-colonial sentiments, the strengthening of African independence movements and the advent of the Cold War (Alden 2001; Alexopoulou and Juif 2017). Locally, factors related to the labour market, including a racialised labour regime that prevented indigenous people from accessing skilled and semi-skilled jobs, forced labour, profiteering from the labour market to South Africa, unjust labour treatment, illegal extension of contracts, underpayment and withholding of wages caused grievances that led some Mozambicans to take up arms against the Portuguese (Alexopoulou and Juif 2017).

Among the Anglophone countries in the SADC region, the Namibian war for liberation and democracy against the South African apartheid government from 1960 to 1990, led by the South West African People's Organisation (SWAPO) (Heuva 2003), provides a case study of the way in which labour market issues contributed to war. SWAPO was formed out of a Namibian labour movement called the Ovamboland People's Congress (OPC) in 1957 and later renamed Ovamboland People's Organisations (OPO) (Mbenzi 2015). After an unsuccessful unarmed struggle against the abusive, exploitative and racist labour market regulation drawn by the apartheid regime, some members of the OPO created the SWAPO to fight against the racist, despotic and exploitative employers of Namibian cheap labour and their backer, the apartheid government of South Africa (Saunders 2003; Mbenzi 2015).

Other strong adversaries of the apartheid regime were the African National Congress (ANC), its allies in unions such as the Congress of South African Trade Unions (COSATU) and its armed wing called *uMkhonto we Sizwe*[2] (Arnold 2007). Despite its involvement in the unionist struggle against racism in the South African labour market, the ANC had abandoned the non-violent struggle to create its armed wing to fight the apartheid government (Arnold 2007; Buhlungu and Tshoaedi 2013).

Angola (1975–2002), Mozambique (1975–1992) and the DRC make up three countries that went through civil wars in the SADC region (1960–1965; 1977, 1978, 1996–1997; 1998–2002) (Saunders 2003; Mbenzi 2015; Arnold 2007; Buhlungu and Tshoaedi 2013). The Angolan and Mozambican civil wars started just after these countries, respectively, gained political independence from Portugal in 1975 (Bekoe 2008). Although there were some labour market problems, these civil wars were more due to power struggles between the MPLA and UNITA in Angola and between FRELIMO and the Mozambican National Resistance (RENAMO) in Mozambique (Arnold 2007; Bekoe 2008).

However, the DRC, a Francophone member state of the SADC, offers two rounds of civil wars: between 1960 and 1964 and between 1998 and 2002, which were directly triggered by conflicts over the labour market. The next chapter will discuss the causes of those wars in detail. In this chapter at hand, we just make a passing reference to the immediate causes of the wars. From 1960 to 1965, the DRC went through a period of post-independence civil wars that were triggered by military personnel's inability to be promoted or rise to higher ranks in the Congolese Army (Inaka 2020).

The foregoing partly highlights the fact that labour market issues produced fertile grounds for all liberation and/or civil wars in Southern Africa. These, in turn, resulted in the poor functioning of labour markets in the affected countries of the region. Although all these countries already had several labour market issues before and during the wars, the point worth reiterating is that their labour markets were often ruined in post-war periods. Hence, post-war reconstructions of the labour markets

[2] The Spear of the Nation in South African Zulu local language.

of these countries were very imperative. But it is curious to note that the SADC, as a regional body, gave little attention to these processes. For the sake of clarity, it is worth briefly recalling the historical background of the SADC in order to understand why it seems to be less involved in post-war labour market reconstruction.

The Origins of SADC and the Apathy Towards Post-war Labour Market Reconstruction in Southern Africa

The SADC has its origins in the Southern African Development Coordination Conference (SADCC), established in Lusaka in 1980 (Afadameh-Adeyemi and Kalula 2010; Van Nieuwkerk 2014). The SADCC was created to foster economic cooperation among its members (Afadameh-Adeyemi and Kalula 2010). It also sought to reduce economic dependence on the apartheid government of South Africa at the time (Afadameh-Adeyemi and Kalula 2010). Despite its lofty goals of regional integration, SADCC had neither a developed institutional structure nor a binding legal framework (Lenz 2012). Moreover, its institutions were decentralised in that, individual states were responsible for the development of particular sectors such as infrastructure, food and security (Afadameh-Adeyemi and Kalula 2010). Nonetheless, the great merit of SADCC is that it succeeded in combating and eradicating colonialism and racism that existed in Southern Africa (Lenz 2012; Van Nieuwkerk 2014).

With the independence of Zimbabwe (1980) and Namibia (1992), the demise of apartheid (1994) and especially the end of the Cold War, SADCC was transformed into SADC in 1992 as a regional organisation with aims based more on economic and political integration in Southern Africa (Schoeman 2002; Evans 2010; Nagar and Malebang 2016). To that end, several policies on regional integration have been drawn since the 1990s (Links 2020). For example, the SADC's blueprint for accelerating and strengthening regional integration, known as the SADC Regional Indicative Strategic Development Plan (RISDP), was first adopted in 2003 (Nagar and Malebang 2016). It provided an overall strategy for the implementation of policies on trade, economic growth, infrastructure development, food security, sustainable human and social development, alleviation of poverty, eradication of HIV/AIDS, improvement of life standards (in terms of gender equality, education, health) and consolidation of regional security in SADC from 2005 to 2020 (Nagar and Malebang 2016; Links 2020).

Although the RISDP set out its priorities, policies and strategies to achieve the above-mentioned objectives, it is curious to note that the first RISDP was not able to achieve all of its intended objectives (Links 2020). In particular, the SADC, having initially emerged from an environment/ characterised by wars, has always considered the consolidation of regional security as one of the priorities for regional integration (Van Nieuwkerk 2014). For this reason, the SADC created a regional security structure and institution called the SADC Organ on Politics, Defence and Security Cooperation (OPDSC) in 1996 (Malan 1998; Zondi 2017). Notwithstanding its objectives to build and maintain regional peace and security, this institution seems to have not yet been able to provide effective solutions in cases of wars or political crises because of conflicting interstate interests, tensions on management operations, and lack of compromise between many countries (Van Nieuwkerk 2014; Nagar and Malebang 2016).

Overall, SADC emerged after the end of several wars in Southern Africa. It is however surprising to note that SADC member states seem to have paid little attention to labour market realities which were among the main causes of wars in the region. Moreover, despite its goals of development, political and economic integration and, especially conflict resolution, peace-making and post-war reconstruction in the SADC region, the body has not yet assisted any member country in post-war labour market reconstruction processes. This inactivity of SADC towards post-conflict labour markets has aroused our curiosity to conduct the present study. Another reason for the study is that even though Data-Bah (2003) and Cramer (2008) repeatedly rail against the paucity of post-war labour market reconstruction studies, the academic community seems slow to respond to these calls. Therefore, our book has its *raison d'etre* in covering this gap in the literature. We do so through a set of methodological, empirical and theoretical tools that we explain next.

Methodological and Theoretical Considerations

Our main objective is to understand the role of a range of actors including, the state, labour market institutions, employers, employees and non-market actors in the reconstruction of post-war labour markets. We focus on Southern Africa as a region, which, as we have attempted to show in this introduction, has gone through various types of war and conflict since the advent of colonialism. Within the SADC region, we narrow down the discussion (in the empirical section of the book) to a member state, the DRC. This enables us to richly engage with qualitative information from the interwoven complexities and multiple factors that define post-war labour market conditions. In turn, it permits the analysis of labour market reconstructions in the SADC region's member states. Our approach finds its basis in the research methods literature, which legitimises it as appropriate for both quantitative and qualitative research (Cronin 2014; Merriam 2009; Yin 2014; Creswell and Creswell 2018). The approach simultaneously enables us to hold a holistic perspective of actual events in the SADC region that include, among others, the behaviours of (small) groups, organisational processes, the growth of industries and international relations (Yin 2014). We are cognisant of the fact that although the majority of the countries in the SADC region went through liberation wars and/or struggles for independence, and were exposed to and served as a theatre of the Cold War (Graham 2010; Lee 2011), the ways in which these experiences occurred in each country vary.

The main objective of the book is built on four specific objectives. The first is to examine the role of the state in reconstruction processes of post-war labour markets. The second and third are to understand the contribution of labour market institutions to the reconstruction of the post-war labour markets and to analyse the participation of the private sector in the remaking of the labour markets. The fourth is to understand the experiences of employees (specifically those employed by companies owned by Indian, South African and Chinese capital) in finding jobs in labour markets that are reconstructing or being reconstructed. We draw specific insights that speak to each of these objectives from the DRC.

We deploy a constructivist methodological approach built on ethnography to achieve the objectives of the study. The ethnographic exercise enabled us to collect detailed data through interviews conducted during the fieldwork for study in Kinshasa over a four-year period, from 2016 to 2019. As there are many actors involved in the reconstruction of post-war labour markets in countries such as the DRC, we deemed it important for the study to generate detailed data from various categories of informants. This allowed for a deeper and further understanding of the contradictions and complexities that prevail in labour markets. We also found this to be consistent with Yin's (2014) recommendation, as it enabled the investigation of complex social units that include multiple variables which are crucial to the understanding of the realities of post-war labour markets. The first category of informants included academics, who studied the making and remaking of the Congolese labour market. The category also included trade unionists, who had extensive experience of historical and current realities of the Congolese labour market. The second category of informants comprised senior executives in public labour market institutions. These senior executives spoke about how their institutions influenced the reconstruction of the Congolese labour market. The third category of respondents included extra-market actors such as religious leaders, politicians, human rights activists, traditional leaders, journalists, academics and unofficial labour lobby groups such as *24 entreprises et subséquentes*.[3] Additionally, we triangulated the information gathered through interviews with some useful data sourced from official government publications, academic publications, newspapers, radio and television stations, online news media sources and archives. While we make reference to the information from the interviews across the book, we present and discuss most of it in the empirical chapters (i.e. Chapters 4 and 5 and especially Chapters 6 and 7). The historical and macro-level data for the book (especially this chapter, Chapters 2–4) drew on the secondary sources of data cited above and is also based on the theoretical foundation of the book.

At the theoretical level, we draw on a combination of concepts and theoretical tools that we found peculiarly applicable and relevant to the DRC and Southern African labour market realities. Among them, we also borrow from Peck's theory of labour market social regulation. The theory argues that labour markets are historically constructed and embedded. Our study, however, goes beyond Peck's position by showing that the Congolese labour market like most of the other SADC member states' markets was particularly historically constructed. It was constructed before war, wrecked during times of wars and reconstructed after wars.

[3] *24 entreprises et subsequentes* is a labour lobby group in the DRC.

This is to say that, labour markets are in perpetual movements of structuration and deterioration, or explosion and restructuration. In simpler terms, labour markets can be made before the occurrence of wars, unmade during wars and remade after wars, and this can happen for a succession of wars. Because Peck's theory has limitations in fully capturing the quintessential labour market realities of post-war African countries, it needs to be complemented by other and various theories on African states to better understand the post-conflict reconstructions of labour markets. An identifiable aspect is that of patrimonialism, a tendency by which the right to rule is ascribed to individual rulers rather than to a ruling administration (Clapham 1985; Thomson 2010). Bratton and Van de Walle's (1994) notion of neopatrimonialism actually further postulates that chief executives of companies maintain authority through personal patronage rather than through ideology or law. There is also Chabal and Daloz's (1999) notion of the 'instrumentalization of disorder', which assumes that rulers take advantage of weak institutions in their countries to satisfy various needs. Added to this, Bayart et al. (1999) and Bayart (2006) suggest that leaders often turn to corruption instead of performing their accredited roles. Lastly, Mbembe (2000) proposes the concept of indirect private government to explain a regime of power that operates to the advantage of strong individuals who mercilessly compete for the concentration of power. We thus found these notions of patrimonialism, neopatrimonialism, instrumentalisation of disorder and politics of the belly to be relevant and important theoretical tools for explaining post-war labour market reconstructions in Southern African states, particularly the DRC.

The Structure of This Book

This book is organised into eight chapters. This introductory chapter provided a diachronic global view of post-war labour market reconstructions, highlighted the gap in the literature on post-war labour market recovery in Southern Africa and revealed the novel empirical and theoretical contribution of the book to studies of labour market recoveries and labour market theories to various disciplines. The chapter also highlighted the theoretical perspectives and methods used and next outlines the contents of the rest of the book.

Chapter 2 briefly discusses the history of the SADC region, to give a background that emphasises five key issues on post-war labour market reconstruction in Southern Africa. First, the fact that countries in the SADC region have been exposed to various and a range of wars that have impacted labour markets. Second, the chapter gives an historical overview of the creation of waged labour markets in SADC countries by emphasising the particularities of each former colonial power over the respective subjugated territories. Third, it explains how the advents of war contributed to changes of/in labour market regimes and to the emergence of labour movements and the acceleration of decolonisation processes in the region. Fourth, the chapter presents the different labour regimes that have existed in post-independent SADC countries. Finally, it analyses the ruining of labour markets in some SADC countries and paves the way for a detailed examination of their reconstruction in subsequent chapters of the book.

Chapter 3 picks up on the discussion from Chapter 2 and proceeds to engage with the definitional and theoretical issues of the book. The chapter provides the lenses through which we analyse the creation, break down and reconstruction of labour markets in countries that have experienced war and proceed to post-war status in a region such as the SADC. Given the focus of the book and that no study of the nature we present herein that focuses on the SADC region has been conducted, the chapter starts by exploring the five basic concepts utilised in the book for this purpose. The concepts in question include, labour market; labour market institutions; the state wherein such institutions exist, break and are reconstituted; power and the international system. The chapter theoretically engages with classical, neoclassical and institutionalist labour market paradigms; labour segmentation theory and Peck's theory of labour market social regulation. It argues that it is only through a combination of these perspectives that post-war labour market reconstruction particularly in member states of the SADC can most aptly be explained. Furthermore, it suggests that deploying Peck's theory of labour market social regulation has particular relevance to the SADC member country on which the book focuses—the DRC—to engage in detailed analysis of the complexities that define the dynamic making, unmaking and remaking of post-war labour markets. The chapter shows that the links manifested in the labour market of the DRC between the past and the present are consistent with Peck's argument that labour markets are historically constructed and embedded.

Chapters 4–7 constitute the empirical section of this book. They narrow down the discussion of post-war labour market reconstructions in the SADC region to the realities that obtain in a member state, which is the DRC. Chapter 4 begins the application of the conceptual and theoretical perspectives laid out in Chapter 3. It applies them to the historical development of the labour market in the DRC. The chapter traces past aspects of the labour market in the DRC that contribute to its current shape. Hence, it analyses the implementation of the forced labour regime in the Congo Free State (1885–1908) and the racialised dual labour market in the Belgian Congo (1908–1960). It then goes on to show how wars in the country balkanised the post-independence labour market (1960–1965) of the DRC and also shows how the country's labour market faced an endless process of regress during the respective eras of President Mobutu (1965–1997) and President Laurent-Désiré Kabila (1997–2001) after the 1996–1997 war. The chapter furthermore examines factors that conditioned and inhibited attempts by the state in the DRC to achieve post-war labour market reconstruction between 2003 and 2019, during President Joseph Kabila's reign (2001–2019). The chapter thus shows that the power-sharing deal brokered during Joseph Kabila's presidential term negatively impacted the reconstruction of the post-war labour market because it hindered reconstruction during the transition period of 2003–2006. Because of this, the chapter argues that despite the DRC adopting a new constitution in 2006 and electing new governments in 2006 and in 2011, the implementation of labour reforms in the country was marked by significant delays at the expense of the labour force. Some statistical information presented later in the chapter supports this argument by emphasising the state's failure to implement post-war labour market reforms and other challenges.

Chapter 5 builds on the discussion in Chapter 4 and continues the focus on the DRC as a member state of the SADC. It analyses the role and contribution of labour market institutions to the process of post-war labour market reform and recovery in the DRC. The chapter examines six public labour market institutions to determine their challenges and evaluate the role they play in the reform of the labour market of the DRC. In so doing, the chapter shows the correlation between the internal weaknesses of labour market institutions and those imposed by factors that are external to the DRC. Chapter 6 continues the analysis of the contribution and role of labour market institutions in post-war labour market reform and recovery but shifts focus away from public labour market institutions. Instead, it examines private labour market institutions including employer organisations and trade unions. Through this, the chapter shows that private institutions too, contribute to remaking the labour market, albeit feebly. In addition to private labour market institutions, Chapter 6 also considers the involvement of non-market actors in those institutions and how the non-market actors weaken the institutions and weaken the contribution of the institutions to post-war labour market reconstruction.

Chapter 7 shifts the focus further away from the state and the private sector and their respective labour institutions. It discusses employees and examines how they integrate into the private sector, particularly in foreign companies owned by other Africans, Indo-Pakistanis, Lebanese, Europeans and Chinese, who invest in the DRC's construction, retail and telecommunication sectors. The chapter also examines the communication that occurs between employers and prospective employees to dissect the strategies that employers adopt in order to inform job seekers and the general public about job vacancies in their companies. It shows that many employers do not follow government procedures when advertising job vacancies. Instead, they use employer recruitment networks and worker supply networks to source for employees. Chapter 7 additionally compares the channels that job seekers use to access information on job opportunities. Based on responses obtained from interviews, it shows that job seekers often use strong social ties rather than weak ties to get jobs, even though these ties are not mutually exclusive in some job search processes. The chapter further discusses the ways in which employees are hired and posted to different positions in companies and shows that only a few respondents could actually get permanent jobs. This is because permanent employment usually comes through employer/worker supply networks. And yet the chapter makes clear that the match between employee profiles and their jobs varies according to skills required in a given sector and/or the fact of belonging to employer/employee networks. The more employees possess desirable skills in companies, the more they hold jobs that correspond with their human capital.

As the final chapter, Chapter 8 provides some concluding remarks to the book. These include the argument that the labour market in the DRC was constructed before the country experienced any war and that, that labour market got wrecked before and/or during the wars the country experienced and was reconstructed after the wars. The chapter also underscores the theoretical contribution of the book and posits that labour markets are not only socially, politically and historically constructed, but are also destructed and reconstructed in certain ways. In other words, labour markets are in the perpetual motion of structuration, deterioration and restructuration. They can be made before wars, unmade before and/ or during wars and remade after wars. These theoretical postulations have empirical implications and can be applied to various extents to the DRC's neighbouring countries and member states of the SADC.

References

Afadameh-Adeyemi, Ashimizo and Evance Kalula. 2010. "SADC at 30: Re-examining the Legal and Institutional Anatomy of the Southern African Development Community." *Monitoring Regional Integration in Southern Africa, Yearbook* 10.

Alden, Chris. 2001. *Mozambique and the Construction of the New African State. From Negotiations to Nation Building*. New York: Palgrave Macmillan.

Alexander, Jocelyn, JoAnn McGregor and Blessing-Miles Tendi. 2017. *The Transnational Histories of Southern African Liberation Movements: An Introduction*. Taylor & Francis.

Alexander, Robert Jackson. 2002. *A History of Organized Labor in Cuba*. Greenwood Publishing Group.

Alexopoulou, Kleoniki and Dácil Juif. 2017. "Colonial State Formation Without Integration: Tax Capacity and Labour Regimes in Portuguese Mozambique (1890s–1970s)." *International Review of Social History* 62(2):215–52.

Amarasuriya, Harini, Canan Gündüz and Markus Mayer. 2009. "Rethinking the Nexus Between Youth, Unemployment and Conflict–Perspectives from Sri Lanka." *Strengthening the Economic Dimensions of Peacebuilding Case Study Series, International Alert*. Retrieved from http://bit.ly/4uXTux.

Arnold, Guy. 2007. *Historical Dictionaries of War, Revolution, and Civil Unrest*. Lanham: The Scarecrow Press.

Barakat, Sultan. 2002. "Setting the Scene for Afghanistan's Reconstruction: The Challenges and Critical Dilemmas." *Third World Quarterly* 23(5):801–16.

Bayart, Jean-François. 2006. *L'Etat en Afrique: la politique du ventre*. Paris: Fayard.

Bayart, Jean-François, Stephen Ellis and Béatrice Hibou. 1999. *The Criminalization of the State in Africa*. Oxford: James Currey.

Beasley, Kenneth W. 2006. "Job Creation in Postconflict Societies." *Federal Publications*:274.

Bekoe, Dorina. 2008. *Implementing Peace Agreements: Lessons from Mozambique, Angola, and Liberia*. New York: Palgrave Macmillan.

Berg, Janine and David Kucera. 2008. "Labour Institutions in the Developing World: Historical and Theoretical Perspectives." Pp. 9–31 in *In Defence of Labour Market Institutions*, edited by J. Berg. London: Palgrave Macmillan.

Binningsbø, Helga Malmin and Kendra Dupuy. 2009. "Using Power-Sharing to Win a War: The Implementation of the Lomé Agreement in Sierra Leone." *Africa Spectrum* 44(3):87–107.

Bratton, Michael and Nicolas Van de Walle. 1994. "Neopatrimonial Regimes and Political Transitions in Africa." *World Politics* 46(4):453–89.

Bray, John. 2005. *International Companies and Post-Conflict Reconstruction*. Washington, DC: World Bank.

Brewer, John. 2010. *Peace Processes: A Sociological Approach*. Cambridge: Polity.

Brewer, John. 2013. "Sociology and Peacebuilding." Pp. 159–70 in *Routledge Handbook of Peacebuilding*. London and New York: Routledge.

Britwum, Akua O 2012. "Labour in African History: Trends and Organisational Forms." Paper presented at the Second ITUC-Africa New Year School, Kisumu, Kenya.

Buhlungu, Sakhela and Malehoko Tshoaedi. 2013. "A Consted Legacy: Organisational and Political Challenges Facing Cosatu." Pp. 1–31 in *COSATU's Contested Legacy: South African Trade Unions in the Second Decade of Democracy*, Vol. 28, edited by S. Buhlungu and M. Tshoaedi. Leiden and Boston Brill.

Bvirindi, Jeremiah. 2019. "Angolan Conflict Analysis: A Lesson for Africa's Emerging Leaders." *International Journal of Research and Innovation in Social Science* 3(7):2454–6186.

Chabal, Patrick and Jean-Pascal Daloz. 1999. *Africa Works: Disorder as Political Instrument (African Issues)*. Oxford: James Currey.

Chomsky, Aviva. 2011. *A History of the Cuban Revolution*. West Sussex: Wiley-Blackwell.

Clapham, Christopher S. 1985. *Private Patronage and Public Power: Political Clientelism in the Modern State*. London: Frances Pinter.

Collier, Paul. 2000. "Rebellion as a Quasi-Criminal Activity." *Journal of Conflict Resolution* 44(6):839–53.

Collier, Paul, ed. 2011. *Wars, Guns and Votes: Democracy in Dangerous Places*. HarperCollins.

Collier, Paul and Anke Hoeffler. 2004. "Greed and Grievance in Civil War." *Oxford Economic Papers* 56(4):563–95.

Cramer, Christopher. 2008. "From Waging War to Peace Work: Labour and Labour Markets." Pp. 121–38 in *Whose Peace?: Critical Perspectives on the Political Economy of Peacebuilding*, edited by M. Pugh, N. Cooper and M. Turner. New York: Palgrave Macmillan.

Cramer, Christopher. 2011. *Unemployment and Participation in Violence. Background Paper for the World Development Report 2011*. Washington, DC: World Bank.

Creswell, John W. and David J. Creswell. 2018. *Research Design: Qualitative, Quantitative, and Mixed Methods Approaches Fifth Edition*. Thousand Oaks, CA: Sage Publications, Inc.
Cronin, C. 2014. "Using Case Study Research as a Rigorous Form of Inquiry." *Nurse Researcher* 21(5):19–27.
Cubitt, P. Christine 2011. "Employment in SIerre Leone: What Happened to Post-Conflict Job Creation?" *African Security Review* 20(1):2–14.
Data-Bah, E. 2003. "Women and Other Gender Concerns in Post-Conflict Reconstruction and Job Promotion Efforts." *Jobs After War. A Critical Challenge in the Peace and Reconstruction Puzzle*: 111–52.
Date-Bah, Eugenia. 2006. *Jobs after War: A Critical Challenge in the Peace and Reconstruction Puzzle*. Geneva: ILO.
Del Castillo, Graciana. 2008. *Rebuilding War-Torn States: The Challenge of Post-Conflict Economic Reconstruction*. Oxford: OUP.
Diumasumbu, Raphaël Mukanga. 2008. "Réforme administrative de la République démocratique du Congo: leçons et travers du passé." *Afrique contemporaine* (3):95–106.
Dunne, Paul. 2003. "Challenges of Armed Conflicts to Jobs and Other Socioeconomic Issues in Africa." Pp. 35–49. in *Jobs After War: A Critical Challenge in the Peace and Reconstruction Puzzle*, edited by E. Date-Bah. Geneva: ILO.
Englebert, Pierre and Denis M. Tull. 2008. "Postconflict Reconstruction in Africa: Flawed Ideas About Failed States." *International Security* 32(4):106–39.
Evans, Jessica. 2010. "The Neoliberal Turn in the SADC." *Nokoko* 1(1):101–29.
Fearon, James D. and David D. Laitin. 2003. "Ethnicity, Insurgency, and Civil War." *American Political Science Review* 97(1):75–90.
Forsberg, Aaron. 2000. *America and the Japanese Miracle: The Cold War Context of Japan's Postwar Economic Revival, 1950–1960*. Chapel Hill and London University of North Carolina Press.
Galtung, Johan. 1969. "Violence, Peace, and Peace Research." *Journal of Peace Research* 6(3):167–91.
Galtung, Johan. 1990. "Cultural Violence." *Journal of Peace Research* 27(3):291–305.
Galtung, Johan. 2004. *Transcend and Transform: An Introduction to Conflict Work*. London: Pluto Press
Galtung, Johan. 2005. *Pax Pacifica: Terrorism, the Pacific Hemisphere, Globalisation and Peace Studies*. Boulder and London: Pluto Pr.
Giddens, Anthony. 1985. *The Nation-State and Violence*. Oxford: University of California Press.
Graham, Matthew. 2010. "Cold War in Southern Africa." *Africa Spectrum* 45(1): 131–39.

Greenberg, Stephen. 2008. "The Next Liberation Struggle: Capitalism, Socialism and Democracy in Southern Africa." *Transformation: Critical Perspectives on Southern Africa* 65(1):195–200.

Guimaraes, Joao da Costa Cabral Andresen. 1992. "The Origins of the Angolan Civil War. International politics and Domestic Political Conflict 1961–1976." London School of Economics and Political Science (United Kingdom).

Gustafsson, Mai Lan. 2010. *War and Shadows: The Haunting of Vietnam*: London: Cornell University Press.

Harrison, Mark. 2002. *Accounting for War: Soviet Production, Employment, and the Defence Burden, 1940–1945*, Vol. 99. Cambridge: Cambridge University Press.

Hartzell, Caroline A and Matthew Hoddie. 2007. *Crafting Peace: Power-Sharing Institutions and the Negotiated Settlement of Civil Wars*. Pennsylvania: Pennsylvania State Press.

Heuva, William. 2003. "Voices in the Liberation Struggle." *Re-Examining Liberation in Namibia*:25.

Inaka, Saint Jose. 2020. "Post-war Labour Market Reconstruction: The Case of the Democratic Republic of the Congo." Ph.D, Sociology, Vol. Pretoria: University of Pretoria Pretoria.

Izzi, Valeria. 2013. "Just Keeping Them Busy? Youth Employment Projects as a Peacebuilding Tool." *International Development Planning Review* 35(2):103–17.

Jarstad, Anna K. 2008. "Power Sharing: Former Enemies in Joint Government." Pp. 105–33 in *From War to Democracy: Dilemmas of Peacebuilding*, edited by A. K. Jarstad and T. D. Sisk. Cambridge: Cambridge University Press.

Joas, Hans and Wolfgang Knöbl. 2012. *War in Social Thought: Hobbes to the Present*. Princeton and Oxford Princeton University Press.

Joes, Anthony James. 2001. *The War for South Viet Nam, 1954-1975*. London: Greenwood Publishing Group.

Kaldor, Mary. 2012. *New and Old Wars: Organised Violence in a Global Era*. Cambridge: Polity

Kalyvas, Stathis N and Laia Balcells. 2010. "International System and Technologies of Rebellion: How the End of the Cold War Shaped Internal Conflict." *American Political Science Review* 104(3):415–29.

Keen, David. 2012. "Greed and Grievance in Civil War." *International Affairs* 88(4):757–77.

Kester, Gérard. 2007. *Trade Unions and Workplace Democracy in Africa*. London and New York: Routledge.

Kevlihan, Rob. 2013. *Aid, Insurgencies and Conflict Transformation: When Greed Is Good*. New York: Routledge.

Krishnamurty, John. 2007. "Employment and Armed Conflict." *Indian Journal of Labour Economics* 50(1):47.

Kwon, Seung Ho and Michael O'Donnell. 2003. *The Cheabol and Labour in Korea: The Development of Management Strategy in Hyundai*. New York and London: Routledge.

Leão, Ana and Martin Rupiya. 2005. "A Military History of the Angolan Armed Forces from the 1960s Onwards—as Told by Former Combatants." Pp. 7–41 in *Evolutions and Revolutions: A Contemporary History of Militaries in Southern Africa*, edited by M. Rupiya. Pretoria: Institute for the Security Studies.

Lee, Christopher J. 2011. "Decolonization of a Special Type: Rethinking Cold War History in Southern Africa." *Krono* 37(1): 6–11.

Lenz, Tobias. 2012. "Spurred Emulation: The EU and Regional Integration in Mercosur and SADC." *West European Politics* 35(1):155–73.

Levitsky, Steven and Lucan A Way. 2010. *Competitive Authoritarianism: Hybrid Regimes After the Cold War*. New York: Cambridge University Press.

Levitt, Jeremy 2012. *Illegal Peace in Africa: An Inquiry into the Legality of Power Sharing with Warlords, Rebels, and Junta*. New York: Cambridge University Press.

Links, Stacey. 2020. "Parameters and Pathways: Agency in the Case of the Southern African Development Community." Pp. 141 in *Global Perspectives on China's Belt and Road Initiative*, edited by F. Schneider. Amsterdam: Amsterdam University Press.

Machado, Barry. 2007. *In Search of a Usable Past: The Marshall Plan and Postwar Reconstruction Today*. Lexington: George C. Marshall Foundation.

Magnusen, Karl O. and Leonardo Rodríguez. 1998. "Cuba, Labor, and Change." *Labor Studies Journal* 23(2):21–40.

Malan, Mark. 1998. *SADC and Sub-regional Security*, Vol. 19. Pretoria ISS.

Malešević, Siniša. 2010a. "How Pacifist Were the Founding Fathers?: War and Violence in Classical Sociology." *European Journal of Social Theory* 13(2):193–212.

Malešević, Siniša. 2010b. *The Sociology of War and Violence*. Cambridge: Cambridge University Press.

Mann, Michael. 1988. *States, War and Capitalism: Studies in Political Sociology*. Oxford: Basil Blackwell.

Mbembe, Achille 2000. *De la postcolonie: essai sur l'imagination politique dans l'Afrique contemporaine*. Paris: Karthala.

Mbenzi, Petrus Angula. 2015. "Revolutionary Songs as a Response to Colonialism in Namibia." Pp. 71 in *Re-Viewing Resistance in Namibian History*, edited by J. Silvester. Windhoek: UNAM Press.

McKenzie, Brian Angus. 2007. *Remaking France: Americanization, Public Diplomacy, and the Marshall Plan*. New York and Oxford: Berghahn Books.

McLeod, Darryl and Maria E Davalos. 2008. "Post-conflict Employment Creation for Stabilization and Poverty Reduction." *UN Development Programme, New York.*

McSorley, Kevin. 2014. "Towards an Embodied Sociology of War." *The Sociological Review* 62(2_suppl):107–28. https://doi.org/10.1111/1467-954X.12194.

Mehler, Andreas. 2009. "Introduction: Power-Sharing in Africa." *Africa Spectrum* 44(3):2–10.

Merriam, S. B. 2009. *Qualitative Research: A Guide to Design and Implementation*. San Francisco, CA: Jossey-Bass.

Nagar, Dawn and Gabriel Malebang. 2016. "Region-Building and Peace-Building in Southern Africa."

Nye Jr, Joseph S. and William A Owens. 1996. "America's Information Edge." *Foreign Affairs* 75:20.

Paris, Roland. 1997. "Peacebuilding and the Limits of Liberal Internationalism." *International Security* 22(2):54-89.

Paris, Roland. 2004. *At War's End: Building Peace After Civil Conflict*. Cambridge: Cambridge University Press.

Peck, Jamie. 1992. "Labor and Agglomeration: Control and Flexibility in Local Labor Markets." *Economic Geography* 68(4):325–47. https://doi.org/10.2307/144023.

Porket, J. 1995. *Unemployment in Capitalist, Communist and Post-communist Economies*. Oxford: Springer.

Saunders, Christopher. 2003. "Liberation and Democracy a Critical Reading of Sam Nujoma's 'Autobiography'." *Re-examining Liberation in Namibia Political Culture Since Independence, Uppsala: Nordic Africa Institute*:87–98.

Schoeman, Maxi. 2002. "From SADCC to SADC and Beyond: The Politics of Economic Integration." Pp. 1–12 in *Congress of the International Economic History Organisation, Buenos Aires*.

Shubin, Vladimir Gennad'evich. 2008. *The Hot "Cold War": The USSR in Southern Africa*. London: Pluto Press.

Simons, Claudia, Franzisca Zanker, Andreas Mehler and Denis M. Tull. 2013. "Power-Sharing in Africa's War Zones: How Important Is the Local Level?" *The Journal of Modern African Studies* 51(4):681–706.

Sorel, Eliot, Pier Carlo Padoan, Co-operation Organisation for Economic and Development. 2008. *The Marshall Plan: Lessons Learned for the 21st Century*. Paris: OECD.

Stewart, Frances. 2008. "Horizontal Inequalities and Conflict: An Introduction and Some Hypotheses." Pp. 3–24 in *Horizontal Inequalities and Conflict: Understanding Group Violence in Multiethnic Societies*, edited by F. Stewart. New York Palgrave Macmilan.

Stewart, Frances. 2012. "Employment Policies and Horizontal Inequalities in Post-Conflict Situations." Pp. 61–83 in *Horizontal Inequalities and Post-Conflict Development*, edited by A. Langer, Frances Stewart and R. Vanugopal. London Springer.

Stewart, Frances. 2015. "Employment in Conflict and Post-Conflict Situations. Background Think Piece for Human Development Report." Vol. New York: UNDP Human Development Report Office.

Stewart, Frances. 2016. *Horizontal Inequalities and Conflict: Understanding Group Violence in Multiethnic Societies*, edited by F. Stewart. New York: Palgrave Macmillan.

Stewart, Frances and Arnim Langer. 2008. "Horizontal Inequalities: Explaining Persistence and Change." Pp. 54–82 in *Horizontal Inequalities and Conflict: Understanding Group Violence in Multiethnic Societies*, edited by F. Stewart. New York Palgrave Macmilan.

Thomson, Alex. 2010. *An Introduction to African Politics*. New York: Routledge.

Tilly, Charles. 1985. "War Making and State Making as Organized Crime." Pp. 35–60 in *Violence: A Reader*, edited by C. Besteman. New York: New York University Press.

Toropov, Brandon. 2000. *Encyclopedia of Cold War Politics*. New York: Infobase Publishing.

Tran, Trung Tin. 2017. "Corporate Social Responsibility (CSR) Situation in Vietnam." Degree Programme in International Business Faculty of Business Administration Lappeenrant Saimaa University of Applied Sciences South Karelia, Finland.

Trefon, Theodore. 2010. "Administrative Obstacles to Reform in the Democratic Republic of Congo." *International Review of Administrative Sciences* 76(4):702–22.

Trefon, Theodore. 2011. *Congo Masquerade: The Political Culture of Aid Inefficiency and Reform Failure*. London: Zed Books Ltd.

Truong, Quyen. 2014. "Vietnamese Re-Education Camps: A Brief History." The Choices Program, Brown University. Retrieved 19 April 2016. (https://www.choices.edu/curriculum-catalog/vietnamese-re-education-camps-brief-history).

Tull, Denis M. and Andreas Mehler. 2005. "The Hidden Costs of Power-Sharing: Reproducing Insurgent Violence in Africa." *African Affairs* 104(416):375–98.

Ugarriza, Juan E. 2009. "Ideologies and Conflict in the Post-Cold War." *International Journal of Conflict Management* 20(1):82–104.

Van Nieuwkerk, Anthoni. 2014. "The Strategic Culture of Foreign and Security Policymaking: Examining the Southern African Development Community." *African Security* 7(1):45–69.

Vuong, Quan Hoang. 2014. "Vietnam's Political Economy: A Discussion on the 1986–2016 Period." Vol. Bruxelles: Centre Emile Bernheim, Université Libre de Bruxelles.
Wasser, Solidelle F. and Michael L Dolfman. 2005. "BLS and the Marshall Plan: The Forgotten Story." *Monthly Labor Review* 128:44.
Webel, Charles and Johan Galtung. 2007. *Handbook of Peace and Conflict Studies*. London and New York: Routledge.
Webster, Edward, Rob Lambert and Andries Beziudenhout. 2008. *Grounding Globalization: Labour in the Age of Insecurity*. Malden, MA: Wiley.
Weigert, Stephen L. 2011. "Guerrilla and Conventional Warfare (1978–90)." Pp. 69–103 in *Angola*. Springer.
Weinstein, Jeremy M. 2005. "Resources and the Information Problem in Rebel Recruitment." *Journal of Conflict Resolution* 49(4):598–624.
World Bank. 2008. Kosovo—Youth in Jeopardy: Being Young, Unemployed, and Poor in Kosovo. Washington, DC: World Bank.
Yin, R. K. 2014. *Case Study Research* (5th ed.). Thousand Oaks, CA: Sage.
Yokoi, Noriko. 2004. *Japan's Postwar Economic Recovery and Anglo-Japanese Relations, 1948–1962*. London and New York: Routledge.
Zondi, Siphamandla. 2017. "Comprehensive and Holistic Human Security for a Post-colonial Southern Africa: A Conceptual Framework." *Strategic Review for Southern Africa* 39(1).

CHAPTER 2

Historical Background to the Reconstruction of Post-war Labour Markets in Southern Africa

No complete list exists of types of war or war terms. This is because different perspectives on and aspects of war exist, depending on such factors as motives, determinants and techniques (Grieves 1977). Some literature, however, distinguishes three types of war. These include absolute war, instrumental war and agonistic fighting. There is also literature that typifies war by highlighting, among others, colonial war; insurgency within which is included wars of independence, wars of liberation, civil war and fratricide; cold war; invasion and proxy war. This chapter (and book for that matter) is, however, not concerned with the typification of war. It rather briefly highlights the conflict and war-affected history of the Southern African Development Community (SADC) region. This is in order to help emphasise some issues that pave the way for the examination of post-war labour market reconstruction. Firstly among these issues is the fact that countries in the region have been exposed to wars that have impacted labour markets. Secondly, the chapter gives an historical overview of the creation of wage labour markets by emphasising the particularities of each former colonial power over respective subjugated territories. Thirdly, it shows that war led to decolonisation and, consequently, changes in labour market regimes. Lastly, the chapter (and book) is not a comparative analysis of post-war labour market regimes but their reconstruction in Southern Africa.

S. J. Inaka et al., *The Reconstruction of Post-War Labour Markets in The Southern African Development Community*,
https://doi.org/10.1007/978-3-031-37357-2_2

Most member states of the SADC have made considerable domestic political and economic changes since the late 1980s and early 1990s. They have embraced neoliberalism and consequently seen changes in the policy landscape. This includes the relationship between capital and labour. The policy shifts have brought major changes in the labour market, labour market institutions and labour practices. The region has, therefore, considerably moved towards an institutionalised and harmonised framework of labour law and policy. This framework promotes cooperation and accommodative interactions between labours, employers and governments based on legal rules, procedures and industrial relations (Takirambudde 1995). It is, however, important to understand the background of this emerging and evolving labour market to inform the way forward for the region, especially if the goal of an economically integrated region is to be achieved. The current and emerging labour market in the region generally builds on dilapidated structures due to war(s). This, therefore, essentially represents a reconstruction of the labour market in the countries of the SADC region.

The British, Germans, Portuguese and Belgians in competition with each other colonised this region from circa the mid-nineteenth century. The British colonised the majority of the nation-states in the region. They envisioned developing the region as an integrated economic area, with the idea that the colonies that surrounded South Africa would supply raw materials and cheap labour to that country in Britain's interest (Olsen 1995). The colonial 'policy of deliberate underdevelopment outside South Africa' (Olsen 1995: 154) which the British pursued coupled with the racist policies of European settlers crafted to dominate and control the African population came to define their reign in the region up to the end of colonialism (Arrighi et al. 2010; Bond et al. 2001; Olsen 1995). These policies and the form of rule also informed the wage labour market that subsequently got established in the region and the respective countries, as this chapter shows.

The chapter shows that marketable work and, particularly, wage labour within the category of work that was defined as a gainful occupation in industrialised countries from the nineteenth century onwards, was forcefully imposed on Africans. To accomplish this, the analysis in the chapter only focuses on selected member countries of the SADC. It gives brief labour-related histories of each country from conquests by colonisers and settlers to the creation of wage labour markets. Even so, we take cognisance of scholars like van der Walt's (2007) argument that it is inadequate to understand the early twentieth-century history of labour in Southern Africa through the lens of the nation-state. He cites a couple of reasons why he considers this problematic. Primarily, it is because in a region such as Southern Africa, it is necessary to avoid 'methodological nationalism' in the analysis of labour history that relies on the nation-state as a unit for analysing labour formations. According to van der Walt, studies that focus on the nation-state as the unit of analysis ignore the fact that international labour markets during the period of colonial rule operated across and beyond Southern Africa and the British Empire. In this way, transnational networks critically shaped transnational working-class movements, which this chapter shows contributed to liberation wars that led to the present evolving/emerging labour market(s) in Southern Africa. This was more important in shaping labour markets in the region than nation-state borders. Elsewhere (Nshimbi et al. 2020; Tshimpaka and Inaka 2020), we too challenge the projection of colonial/post-colonial borders on not only the imperial period but contemporary Southern Africa too. They limit the ability to understand a lot of both historical and contemporary socioeconomic and cultural realities in the region. Still, we compartmentalise the discussion of the background to the reconstruction of labour markets in SADC in this chapter. The group of countries we discuss fell under the rule of different colonisers. Yet, the chapter shows that despite the particularities of each former colonial power, colonial rule ultimately led to somewhat similar outcomes insofar as concerns the emerging/contemporary labour markets in the SADC region. African workers in the colonies rallied around the injustices to which they were subjected by colonisers, ultimately leading to political liberation. Similarities and differences existed among British, Portuguese and Belgian colonisers and settlers in the territories they respectively occupied and colonised and, accordingly, impacted the particular histories of these countries. Chapters 4–7 of this book engage in a detailed empirical examination of the reconstruction of the labour market in the Democratic Republic of the Congo (DRC), a former Belgian colony. The following sections in this chapter discuss the labour market histories of selected SADC countries under British and Portuguese colonisers and settlers, highlighting the differences and similarities in the particular histories of the respective countries.

Wage Labour Markets in British Colonies of Southern Africa

South Africa

The South African labour market is racialised and strongly marked by inequalities that are a legacy of the country's past, dating back to the period before the advent of colonialism (Feinstein 2005; Iliffe 1999; Maqutu 2020; Terreblanche et al. 2002). Some scholars use the apartheid regime as an historical marker of the origins of a labour market that racially discriminates against African workers in the country (Barchiesi 2007; Grün 2004; Oosthuizen and Bhorat 2005). A substantial amount of literature on the creation of these labour markets, however, suggests that this discrimination predates the apartheid era (Breckenridge 2008; Feinstein 2005; Maqutu 2020; Terreblanche et al. 2002).

The regulations that define South Africa's labour market had their origins in the Cape Colony, Natal, Transvaal and the Orange Free State. This was before the creation of the Union of South Africa in 1910. Racially discriminatory labour legislation against Africans based on master–servant relationships, vagrancy, pass laws and labour migration shaped the country's labour market. The law also imposed poll and hut taxes on Africans, along with the policy of land dispossession that left them without any means of subsistence. This all forced the Africans into the colonial labour market, often against their will. The combination of racist labour legislation and land dispossession shaped South Africa's racist labour market from inception until the end of apartheid. The racist labour market was a source of the socio-political conflicts, civil strife and wars in the country before the end of apartheid. The following key historical facts help shed more light on this labour market.

The creation of the South African labour market began in the Cape Colony, which was under Dutch East India Company rule in the 1710s. Roman-Dutch law formed the basis of the labour legislation of that time (Hay and Craven 2005; Swiegers 2014). These laws, especially the pass laws and indentured labour acts, sought to use indigenous Khoikhoi labour to supplement the labour provided by slaves and prevent them from deserting their employers.

When English settlers arrived in South Africa, they introduced new laws based on British anti-slave labour market legislation. The country experienced scarcity in the workforce after slavery was abolished in 1807.

The settlers then resorted to labour market policies that forced the local population into the 'waged' labour market to address the labour shortage. The policies in question involved the use of internal sources of labour, apprenticeship and labour migration.

Firstly, the use of internal sources of labour involved the Khoikhoi, the San and prize slaves. The English settlers promulgated this on 1 November 1809 through the Caledon Code on 'Contracts of Hire between the Inhabitants [white people] of [the] ... Colony and Hottentots [i.e., indigenous populations that included the Khoikhoi and San]' in the Cape Colony (Breckenridge 2008; Dooling 2005; Maqutu 2020). The code came into effect after the British had overcome the Burghers (Boers, Afrikaners) in the Cape and abolished slave labour and the slave trade in 1807. It introduced formal waged labour by institutionalising written labour contracts that defined the relationship between white employers, who it termed masters, and African workers, who it called servants (Breckenridge 2008). While the code seemed to be a remedy for unregulated and unpaid labour, it privatised the subjugation of slave labour workers (Breckenridge 2008) and thereby made them the property of their masters. Article 16 of the code required African workers to obtain passes from their employers or magistrates in order to move around the country (Maqutu 2020), failure to which they were considered vagabonds, treated as such and severely punished. The code, therefore, bound African servants to the service of European employers, whose written consent (passes) the Africans needed to leave employers' premises. The code further required them to obtain passes from the district authorities in order to move around anywhere in South Africa. The pass regulations enforced close surveillance of the movements of Africans within the colony, limiting them to their registered 'fixed abodes'.

The Burghers (Boers/trekkers) somewhat adopted similar labour market polices in the Zuid-Afrikaansche Republiek (ZAR) or Transvaal and the Orange Free State, although they went a step further by using forced African labour (Breckenridge 2008; Dooling 2005; Iliffe 1999; Maqutu 2020). The Master and Servant Contract Act (1856), therefore, essentially constituted a disguised form of forced labour market. It followed the formation of two independent republics in the area after the great trek and the Sand River Convention of 1852. In the Great Trek, Boers from the Cape colony migrated into the interior of South Africa (Breckenridge 2008; Maqutu 2020; Swiegers 2014; Worden 2011), as they sought to move away from British domination in the Cape. Once settled in the Transvaal and Free State regions, they founded the said republics as enacted by the Sand River Convention of 1852. In the convention, the British government committed to non-interference in the internal affairs of Boer republics. The Boers had agreed to abandon slavery and form their republic in the north of the Vaal River. Yet, while British settlers established a sort of South African serfdom in the Cape and in Natal, the Boers often used fierce forced labour practices against African natives (Breckenridge 2008; Maqutu 2020).

Secondly, apprenticeship and the prize slaves. Apprenticeship had its roots in the apprenticeship systems that had already existed in the Cape Colony under Dutch rule. The British restructured it by adopting forms of the craft and parochial apprenticeship systems used in Britain in the sixteenth century (Swiegers 2014). Drawing on the Caledon Code, the Cape colony promulged the practice of apprenticeship for African children on 23 April 1812 (Maqutu 2020). By the 1820s, these apprenticeship systems had been adopted in the Cape, Natal, the Orange Free State and the Transvaal (Breckenridge 2008; Swiegers 2014). Apprenticeship affected 'abandoned children' aged between 8 and 18 years old. Further, children born to parents employed by a farmer and were, therefore, dependent on the employer could be apprenticed to that employer. As parental approval or a child's consent was not needed, apprenticeship contracts were signed by employers and magistrates and included information on mutual obligations between apprentices and employers (Swiegers 2014). Therefore, many Khoikhoi children experienced the apprenticeship system in the Cape. The apprenticeship system in the Transvaal was often criticised for its closeness to forced labour and even slavery, as many children there were captured after their parents were killed (Breckenridge 2008; Swiegers 2014). Another source of labour in the Cape Colony included the importation of slaves captured on slave ships. The former slaves were referred to as 'prize slaves' and were usually apprenticed for a period of 14–21 years (Swiegers 2014).

Thirdly, was the use of migrant workers including indentured migrant workers, skilled migrant workers and unskilled African migrant workers. English settlers imported Indians for indentured labour on the sugar cane plantations in the Natal colony from 1860 to 1911 (Harris 2013; Vahed 2019). The Indians were made to sign five-year work contracts, which could be renewed for a further five years (Breckenridge 2008; Vahed 2019). At the end of these contracts, they had the option of returning to India or remaining in South Africa. Although officially invited by the English colonial authorities, the Indian migrant workers were subjected to inhumane treatment, xenophobia and even slavish labour (Harris 2013; Vahed 2019). This made them demand for their rights and engage in revolts and public demonstrations (Vahed 2019). Some authors explain how Leung Quinn Mahamat Ghandi, for instance, defended the rights of these workers and that he not only participated in public demonstrations in Natal in 1913 but repeatedly called for the abolition of the indentured system too (Harris 2013).

Chinese were brought too and indentured in diamond and gold mines of the Transvaal colony (Harris 2013). According to Harris (2013), the use of Chinese indentured labour started after the successive discoveries of rich diamond deposits in Kimberly in 1888 and gold in the Transvaal in 1888. The discoveries contributed to the development of manufacturing and agricultural industries, which increased labour requirements (Swiegers 2014). The need for a large workforce increased because African labour in the post-Anglo-Boer war was in short supply (Bonner et al. 2007; Denoon 1967). Because the English colonialists had failed to import African migrant workers from Central and East Africa, they resorted to importing indentured Chinese workers (Denoon 1967; Harris 2013). The 1904 Transvaal Labour Importation Ordinance shows that the Chinese migrant workers signed three-year work contracts, renewable for a further three years after which they had to be sent back to China (Denoon 1967; Harris 2013; Swiegers 2014). Furthermore, they could only be employed in the Witwatersrand gold mines as 'unskilled miners'. They were confined to compounds built next to the mines where they worked and had to obtain passes to leave the areas for specified periods of time and not exceeding 48 hours. The Chinese were subjected to employment conditions that resembled semi-slavery conditions unlike the Indians, who were British subjects and could claim certain labour rights. They were also subjected to anti-Asian xenophobia; accused of being unfair competitors; having an excessively low standard of living and introducing opium consumption, gambling and many other vices into the moderate, stable and honest population of Johannesburg.

Another group of migrant workers to South Africa included skilled workers, the vast majority of whom came from England and some from the USA, Australia, New Zealand, Italy, Greece and Canada. They worked in mining and other industries following the discovery of diamond deposits in 1867 (Brunger 2003; Swiegers 2014; Willcox 1931). Before their arrival, strikes or work boycotts were offences under the Master–Servant Relations Act (Swiegers 2014). However, their presence changed the situation and the South African labour market, as they introduced trade unionism in South Africa (Arkin 1960; Harris 1991; Ncube 1985). They created several unions having been inspired by English unions. The unions they formed and industrial sectors included, among others, the Artisans and Engine Drivers' Protection Association (1883), the Witwatersrand Mine Workers' Union (1897), the Rand Mine Workers' Union (1898) and the South African Drivers' Firemen's Association (1898) in mining; the South African Operative Mason's Society (1896) and the White Operative Plaster's Society (1906) in construction; the Eastern Province Typographical Society (1883) and the South African Typographical Union (1897) in printing and the Amalgamated Society of Railway and Harbour Servants (1909) in public sector (Swiegers 2014). These unions aimed to protect only the interests of white migrant workers, when they already enjoyed the best working conditions and high wages (Swiegers 2014). They prevented unskilled African workers from acquiring the required skills and from being promoted to positions held by the skilled migrant workers (Swiegers 2014). In short, they established a unionism based on racial discrimination against non-white employees (Harris 1991; Ncube 1985; Swiegers 2014).

Finally, immigrants from other parts of Africa were also imported to work alongside the Africans in South Africa. The vast majority of these workers came from Southern African countries including, Zimbabwe, Mozambique, Lesotho, Botswana, Malawi and Namibia (Landau et al. 2015; Moodie and Ndatshe 1994). These workers suffered all the disadvantages of a labour market based on racial discrimination in favour of white workers. Their wages were low in comparison to those of white workers; they suffered the negative effects of pass laws and confinement to compounds and were not allowed to organise in trade unions (Bezuidenhout and Buhlungu 2011; Maqutu 2020; Webster et al. 2008, Wood and Harcourt 1998). All their attempts to demand labour rights were violently suppressed. Worse still, they were excluded from political participation and the Africans in South Africa considered these other Africans second-class citizens (Landau, Kalula and Fenwick 2015; Maqutu 2020).

Overall, all the mechanisms of exploitation of African labour for the benefit of white capitalists and the colonial state offered no opportunities for the socioeconomic and political development of Africans. These labour problems gave rise to strikes, social tensions, threats of popular uprisings and armed resistance among Africans.

Zimbabwe

The creation of the wage labour market in South African has some similarities with that of Zimbabwe, since both were British colonies. The British government supported John Cecile Rhode's British South Africa Company (BSAC) exploits in Southern Rhodesia (now Zimbabwe) (Malaba 1980; Mlambo 2014). The British army waged Rhodes' wars against the Ndebele and Shona people in that country in the late 1890s. After appropriating all of the Southern Rhodesian territory or the one that now makes up Zimbabwe in 1897 (Henderson 1972; Rennie 1978; van der Walt 2007), BSAC operated as a state-owned company and was reluctant to implement labour legislation that would provide acceptable working conditions for Africans (Malaba 1980; Mlambo 2014; Raftopoulos and Mlambo 2009; Van Onselen 1973). The company established a forced labour regime wherein workers were forcibly recruited from rural areas and deployed to the mines (Van Onselen 1973).

The colonial state later legalised this system through the Compulsory Native Labour Act, 1942. The state used the act to empower itself to conscript 18–45-year-old African males and supervise the labour in a form of forced labour distinguished from the compulsory type that existed in the other British colonies of Northern Rhodesia and Kenya (Johnson 1992). According to Johnson (1992), the state used the act as a coercive instrument for bypassing labour market forces to secure labour for English settlers and farmers.

This then was the context in which English settlers under the auspices of the state created a dual labour market defined by racial inequalities in Zimbabwe. In this way the settlers in Southern Rhodesia or Zimbabwe did as they had done in South Africa (Rennie 1978). They dispossessed Africans of land and placed them in Bantu reserves (Henderson 1972; Malaba 1980; Rennie 1978). They also imposed head, hut and poll taxes on the Africans to force them to work for private European employers and the colonial state (Malaba 1980).

As in the case of South Africa, the colonial state in Southern Rhodesian had passed the Master and Servant Act of 1901 too, which made it a criminal offence to violate labour contracts, and enacted a Pass Law to control the mobility of Africans (Malaba 1980; Rennie 1978; Van Onselen 1973).

The state had also created the Rhodesia Native Labour Bureau (RNLB) in 1903 which offered cheap labour contracts to Zimbabwean workers through somewhat unofficial methods and, subsequently, recruited, transported and sold the labour to farms or mines (Johnson 1992; Malaba 1980; Rennie 1978). RNLB institutionalised such an inhumane system for Africans (locally referred to as *chibaro* or slave-like labour) that the colonial office had to disbanded it on the grounds of bad reputation (Johnson 1992).

Still, Zimbabweans were forced to accept meagre wages, were underemployed, exploited, abused and subjected to various forms of injustice in private companies and public institutions because they were dispossessed of the land they used for subsistence farming (Malaba 1980; Rennie 1978). This was compounded by the struggle they experienced to generate income to pay taxes (Malaba 1980, Van Onselen 1973). Some Zimbabweans adopted a couple of strategies to avoid being coerced into the capitalist system of production, which Arrighi (Arrighi 1967 as cited in Chinguno 2014) referred to as 'stimulation and strangulation'. Some sold their livestock to meet the tax obligations and avoid working in the mines. Others left Zimbabwe and migrated to the South African Witwatersrand industrial complex, which was the most lucrative labour market in the region (Rennie 1978; Van Onselen 1973).

Employers on their part had a serious shortage of labour. This was despite recruiting migrant labour from neighbouring Nyasaland (now Malawi), Bechuanaland (now Botswana) and Mozambique (Henderson 1972; Johnson 1992; Scott 1954). The workers, firstly, preferred migrating to South African or the Copperbelt mining industries in Northern Rhodesia (now Zambia), where they got better pay and were treated much better than in South Rhodesian. Secondly, the Second World War increased the demand for labour to contribute to the war effort and the economic needs of Southern Rhodesia. Employers lobbied the colonial state to recruit and use Zimbabweans as forced labour and gangs of workers, and headmen and traditional chiefs were, therefore, used to recruit labour. This practice continued after the end of the war. The colonial state only stopped the practice in late 1946, after Zimbabweans, and particularly, Chief Mangwende complained (Johnson 1992).

Still, Africans were deprived of labour rights and worked in deplorable conditions, while white workers enjoyed better treatment with the support of the colonial state (Chinguno 2014; Malaba 1980). Racial inequality at all levels in the workplace frustrated the Africans and raised social tensions between them and English settlers (Van Onselen 1973). '[T]he 1927 strike at the Shamva mine…the first industrial action ever seen in the country' (Van Onselen 1973: 237) provides an example of the consequences of racial inequality and the inhumane treatment of Africans in the workplace in the then Southern Rhodesia. Several labour protests, boycotts and unauthorised strikes besides the Shamva incident are recorded in the country's colonial history (Chinguno 2014).

However, the colonial state would come to address such workers' actions through the Compulsory Native Labour Act and regulations. The regulations provided for the prosecution of the workers for negligence of work, resistance or refusal to work (Van Onselen 1973). And despite the rise of African labour struggles influenced and spurred by movements such as the Industrial and Commercial Workers' Union (Van der Walt 2007; Van Onselen 1973), the colonial state countered such movements by enacting legislation like the Industrial and Conciliation Act, 1937, which only recognised white trade unions and collective bargaining (Chinguno 2014).

At the peak of the harsh treatment and poor conditions in which labour operated across Zimbabwe, and given that strike action became insufficient/ineffective in resolving the problems faced by labour, Joshua Nkomo formed the first nationalist political movement, the African National Congress (ANC) of Rhodesia in 1957 to fight racism and for political independence (Chinguno 2014). What followed was that, some Southern Rhodesians under Nkomo's leadership as well as under Robert Mugabe of the Zimbabwe African National Union-Patriotic Front (ZANU-PF) took up arms to free the country from colonialism. The ensuing war would eventually lead to Zimbabwe's independence and a change in the country's labour market regime.

Namibia

Two imperial states, Germany (from 1884 to 1915) and South Africa (1915–1990), successively shaped the creation of wage labour in Namibia. While the Germans only established a form of embryonic wage labour market, it was the South Africans who instead developed a purely racialised wage labour market from 1924 to 1990.

Franz Adolf Luderitz's company was the first outside entity in Namibia before the German Empire established a wage labour market there

(Werner 1993). The German empire supported the businessman, who used underhand methods and tricked Namibian traditional authorities into signing over land ownership to him, to establish a trading post in South West Africa (now Namibia) in 1881. Luderitz, however, failed to exploit diamonds in Namibia and ended up selling his land to the German Empire. This marked the official start of the German colonisation of Namibia in 1884.

German colonial authorities and concessionary companies were so fierce against Namibians to the point of committing genocide (especially of the Herero and Namaqua people) from 1904 to 1908. The crime seems to have a resistance of German colonial authority and forced labour on colonial commercial farms in inhumane conditions (Cooper 2007; Moorsom 1977). The genocide led to labour shortages, which in turn made the German colonial authorities turn to Angola for wage and migrant labour from 1910 to 1914 (Moorsom 1977). This initial wage labour market regime was, however, interrupted/disrupted by the First World War and, subsequently, South African's occupation of Namibia in 1915 (Cooper 1999; Fenwick 2006).

South African settlers implemented racialised labour market regime almost the same as the one that existed in South Africa (Botes 2013; Cooper 2007; Fenwick 2006; Werner 1993). That is, they set up labour market legislation, institutions, policies and practices in Namibia similar to those of South Africa. Consequently, the Namibian Masters and Servants Act of 1920, the Land Settlement Act of 1920, the Indigenous Administration Act of 1922, the Contract of Employment Act of 1925, the Prohibited Areas Act of 1928 and the Urban Indigenous Movement Act of 1951 forced indigenous Namibians into the country's wage labour market. All these laws provided cheap labour to a racialised labour market (Fenwick 2006), grounded on the labour contract system (Botes 2013; Cooper 1999; Moorsom 1977).

To implement the contract labour system in Namibia, the South African authorities created some labour market institutions that served as employment agencies through which Africans sold their labour and white employers could obtain these workers' services (Cooper 1999). In 1925, the South African authorities established two institutions: the Southern Labour Organisation (SLO) and the Northern Labour Organisation (NLO) (Likuwa 2015). SLO recruited and transported workers to their employers, mainly in diamond mines while NLO recruited workers for other mines and companies (Bauer 2002). These two institutions not only formally structured the Namibian wage labour market but also significantly contributed to racial segregation in Namibia from 1924 to 1943 (Quinn 2020). A large body of literature shows that the two institutions had a bad reputation for mistreating Namibian workers (Cooper 1999). They deteriorated Namibian societal traditions by developing a wage labour market and economy (Cooper 1999; Moorsom 1977). They undermined the labour rights of Namibian workers as they did not accord the workers any right to decide or negotiate types of employment, employers, and salaries (Cooper 1999; Likuwa and Shiweda 2017). The institutions did not tolerate workers breaking labour contracts even though employers could not ensure their welfare (Botes 2013; Fenwick 2006; Likuwa and Shiweda 2017). Hence, the way in which these institutions regulated the labour contract system was regarded as a euphemism for modern labour-slavery (Likuwa and Shiweda 2017).

Given the increase of antagonism against these two institutions over their economic interests and legal prerogatives in the labour market, the South African government created the South West African Native Labour Association (SWANLA) as the unique official employment agency for regulating the contract labour system between African employees and white employers in 1943 (Botes 2013; Cooper 2007). SWANLA often transported Namibian-recruited employees to their employers and provided each worker with a blanket, shirt and pair of khaki shorts. These were the only clothes an employee wore during his contract, which frequently lasted one year (Botes 2013; Cooper 1999; Fenwick 2006). The workers lived in confined, uncomfortable and compacted labour camps where 10–15 men shared a room. They were in complete isolation as they were prohibited from going home or receiving visitors (Botes 2013; Moorsom 1977). The workers had long working hours under despotic labour control and draconian forms of punishments by their employers and were also underpaid and ununionised (Botes 2013; Cooper 1999; Fenwick 2006; Moorsom 1977).

Despite SWANLA supplying white employers with cheap labour, Namibia continued experiencing labour shortages as a fundamental problem of the contract labour market regime. Here too, South African colonial authorities turned to migrant workers. They used SWANLA to recruit Angolan, Basuto, Batswana and South African migrant workers (Cooper 1999). This process was also characterised by racial labour market segregation. While the SWANLA and employers mistreated African migrant workers as they had done indigenous Namibians, South

African colonial authorities in Namibia gave more privileges to poor South African white people and Boers who had returned from exile in Angola. They gave them land or better job positions or even finance from the South African government to run businesses in Namibia (Werner 1993).

Some Namibians from Ovamboland migrated to Angola for work because of the repressive labour legislation and totalitarian regulation of the labour market regime in Namibia. According to Keese (2015), this surprised the South African colonial authorities because Namibians received better pay in their country than in Angola. After several investigations, South African colonial authorities realised that these Namibians preferred being underpaid in Angola because the working conditions were better there (Botes 2013). The racialised labour market, abuse of labour by SWANLA and impacts on workers' lives led the South West African People's Organisation (SWAPO) and all of society into waging a war against the apartheid regime that lasted from 1966 to 1990 (Botes 2013; Cooper 2007; Keese 2012). The war led to independence and the collapse of the apartheid labour market regime.

Overall, the countries analysed above show that the patterns of wage labour market creation in the former British colonies of Southern Africa are remarkably similar. The practice of dispossessing Africans of land, the introduction of head and other taxes and the enactment and implementation of racialised labour legislation forced the African to seek wage labour with the settlers and to accept work characterised by poor conditions. This situation was worse under racist political regimes. The exploited people responded to this by engaging in the struggle for political independence. Yet, the experiences in former British colonies were not that different from the way in which wage labour markets were created in former Portuguese colonies.

Wage Labour Markets in Portuguese Colonies of Southern Africa

Portuguese settlers in the Southern African region were somewhat late in creating wage labour markets in Angola and Mozambique but, simultaneously, maintained long-standing forced labour practices in both countries (Alexopoulou 2018; de Grassi 2018; Keese 2012, 2015; Vos 2014; Walter 1973). Coercive (*compelido*) and punitive (*correcional*) forced labour regimes defined the labour markets in Mozambique and Angola from the time slave labour ended in the nineteenth century up to 1962

(Alexopoulou 2018; Keese 2013). Portuguese settlers did not establish uniform wage labour markets across all regions in their colonies, however.

Mozambique

Three different labour market regimes existed in Mozambique, for example (Alexopoulou 2018). According to Alexopoulou (2018), many Mozambicans were used as forced labourers in the northern region, where they worked for low wages in a chartered company called the Niassa Company. In the central region of the country, workers were also forced to work for meagre wages in both the public sector and in two private companies, Mozambique Company and Zambezi Company. The southern region, however, was used by the colonial state as a kind of labour reserve from which it generated income by exporting Mozambican labour. The colonial authorities set up institutions and laws to regulate the lucrative export of labour (Alexopoulou 2018; Paton 1994). The Mozambican Curator in South Africa, for example, regularised documents for undocumented Mozambican migrant workers in that country, collected taxes from the workers on behalf of the Mozambican colonial government and collaborated with South Africa's WNLA on matters that concerned the recruitment of Mozambican workers for South African businesses (Alexopoulou 2018). Mozambican migrant workers themselves saw the South African Rand Gold mines as an 'eldorado' to earn money and as an alternative to evade forced labour in their own country (Alexopoulou 2018).

Portuguese settlers also dispossessed land from Africans, enacted a labour code for them and imposed a tax on them to force them into the wage labour market (Alexopoulou 2018; O'Laughlin 2002). Actually, Mozambicans who failed to pay the hut, poll or indigenous taxes were conscripted into forced labour by colonial officials (Alexopoulou 2018; Keese 2013). Colonial authorities used traditional chiefs (*regulos*) and administrative assistants (*cabos*) as tax collectors and labour recruiters to enforce this forced labour policy.

Mozambican workers in both private and public organisations occupied very low professional positions while the Portuguese settlers placed themselves and their compatriots in superior and lucrative positions. The gross racial inequalities, occupational abuse, persistence of the forced labour regime and refusal to grant Mozambicans political independence constituted grievances that prompted FRELIMO to engage in armed struggle against the Portuguese and liberate Mozambique in 1975. The attainment of political independence led to a change in the labour market regime in Mozambique.

Angola

Angola shares similarities with Mozambique in the way the wage labour market was constructed. The Portuguese settlers in Angola were pressured into ending the forced labour market regime in 1961 by the international community, particularly the International Labour Organization (ILO), the United Nations (UN) and former colonial powers such as France and Britain (Alexopoulou 2018; Vos 2014; Walter 1973).

Prior to this, however, Portuguese settlers exploited Angolans as forced labourers in agriculture and the public service (Ball 2005; Walter 1973). This was despite the settlers enacting the Labour Code in 1910. The settlers used traditional chiefs (*regulos*) and administrative assistants in the public sector (chefes do posto) at the bottom of occupational hierarchies to recruit indigenous Angolans into forced labour (Walter 1973). Angolan workers were generally frustrated that even illiterate Portuguese peasants occupied higher positions than better educated Angolans (Walter 1973). Another source of frustration for Angolans was their forced exploitation of coffee plantations in northern Angola or cocoa plantations to which they were exported in Sao Tome and Principe (Kagan-Guthrie 2011; Walter 1973). While some Angolans went into exile in the DRC in order to avoid abuse by the Portuguese colonialists (Birmingham 1982; Keese 2012), others stayed home and engaged in armed revolts against the abusers (Keese 2013; Walter 1973). This eventually culminated in armed anti-colonial movements including the People's Movement for the Liberation of Angola (MPLA), National Liberation Front of Angola (FNLA) and the National Union for the Total Independence of Angola (UNITA). The armed struggles in which these groups engaged led to the end of Portuguese colonial rule in Angola in 1975. The independence subsequently resulted in a change to the labour market regime in the country.

The forgoing discussion shows that colonialists used more or less the same procedure to create wage labour markets in SADC countries. Land dispossessions, institutionalisation of head, poll and hut taxes and racialised labour laws compelled Africans to seek wage labour from the settlers and colonial administrations. Africans were deprived of all political and civic rights, forbidden to organise in trade unions and limited in their access to higher education so that the colonisers could better exploit them and reproduce these racial labour market regimes. In response, Africans fought for their freedoms against settlers and colonial regime. The wars led to the end of colonisation followed by changes in the labour market regimes of these countries when they emerged from the conflicts.

The Impacts of War on Labour Markets

A couple of general impacts of war on labour markets can be highlighted from the foregoing and the broader literature on impacts of war. While the armies engaged in war suffer casualties, its most grave consequence is the loss of life. War also has devastating impacts on civilian populations whose lives are spared, especially women and children. It also directly impacts on population displacement. The people who are displaced most likely suffer socioeconomic vulnerabilities. Kondylis (2010) empirically shows that displacement induced by the Bosnia and Herzegovina war had negative impacts on labour market outcomes vis-à-vis access to employment for Bosnian women and men. This was despite that over time, the affected populations seemed to have significantly assimilated into the labour market in that region. War also, obviously, negatively affects the economic output of affected countries and causes damage to physical infrastructure. This is financially costly, as it means that a lot of funds are required for rehabilitation and reconstruction.

The consequences of war on labour markets and implications for post-war reconstruction can manifest themselves in labour institutions and the supply and demand for labour. War disrupts the operation of labour institutions. During war and in its aftermath, labour legislation and regulations may be suspended and/or poorly enforced. Physical infrastructure like roads and employment facilities like factories are destroyed or in a state of disrepair. He destruction of production facilities due to war also implies the destruction of employment opportunities.

War causes death. Where this particularly constitutes economically active individuals, it means the loss of productive labour and negatively impacts its supply. The displacement of people due to war too, means that productive populations are forced to leave the areas and countries of usual residence. This may impact on the supply of labour in the places/countries they leave. Some people within the labour force that is lost to war through death or displacement may possess special or peculiar qualifications. This, then, implies that the losses in the labour force include the loss of quality labour supply to the labour market.

Lastly, war may lead to a rechannelling and repurposing of labour into activities that support the war cause. In such instances, able-bodied men and, unfortunately, children may be conscripted or forced into serving as combatants (Vindevogel et al. 2011). Studies conducted in countries like the DRC and Uganda indicate that employment opportunities also open up in ventures that reproduce war, to generate finance for war, and where children are conscripted into such ventures or as child combatants (Carpenter 2012; Eichstaedt 2011; Samset 2002).

Post-war Labour Market Regimes in Southern Africa

Africans in South Africa, Namibia, Zimbabwe, Mozambique and Angola fought their colonisers to liberate themselves from racism and colonisation. The wars of liberation ended racial and oppressive political regimes and changed the labour market regimes established by colonisers. We briefly discuss these post-war labour market changes as follows.

Angola and Mozambique were the first of the five countries discussed in this chapter to free themselves from colonialism by force of arms in 1975. Upon attaining independence, the new governments in both countries established socialist-oriented labour markets. The governments of both countries naturally opted for state-led economies and labour markets, as socialist or communist countries supported their struggles against Portuguese colonialists. To implement these socialist labour market regimes, both countries nationalised the economy and certain private companies and enacted labour laws based on principles of equal exchange. To remedy the scarcity of skilled labour, they imported foreign experts to help administer their countries and train cadres. They sent citizens abroad for training as well. In Angola, for example, Cuba sent experts to the army and other public services.

Both Mozambique and Angola, however, faced challenges in effectively implementing their socialist-oriented labour market policies as they each descended into the post-independence civil wars. The Cold War in the immediate post-independence periods of these countries strongly shaped the civil wars there. The civil wars in both countries ended in the period after the end of the Cold War. This was the time when the winds of neoliberal reform and democratisation were sweeping across the Southern African region (Tshimpaka and Nshimbi 2022). Both countries found themselves in a region that embraced an approach to regional integration based on neoliberal policies. Therefore, they too as members of the regional group, SADC, had no choice but to implement capitalist labour market regimes and accept neoliberal economic and labour reforms.

Nevertheless, studies have highlighted the difference in political and economic governance and labour market regulation between post-war Mozambique and Angola (Alden 2007; Chabal 2001; Kyle 2014; Pérez Niño and Le Billon 2014). While Mozambique has made efforts to grow its economy and improve its labour market, Angola, despite its natural resources such as oil and diamonds, has paradoxically experienced economic regression in the post-war period. Regardless of these differences, both countries, like several other SADC countries, face subcontracting, precarious employment, poor working conditions, rising unemployment and union erosion.

Unlike the former Portuguese colonies which experienced both liberation and civil wars, the former British colonies only experienced liberation wars against colonialism/apartheid. As an illustration, Zimbabwe was ruled by Robert Mugabe's ZANU-PF after the war against colonialism from 1980 to 2017. The Mugabe regime failed to meet the aspirations of African workers and the Zimbabwean people who hoped for the improvement of the labour market and the whole Zimbabwean society after independence. Contrary to this hope, Mugabe's dictatorial regime only deteriorated the economic fabric of Zimbabwe through poor governance, corruption and ill-conceived and brutally executed land reforms. This has led to impoverishment, political and socioeconomic crises in Zimbabwe. Even after toppling Mugabe from office, no evidence exists so far that suggests improvements in the Zimbabwean labour market. The post-war labour market in Zimbabwe, therefore, essentially remains poorly regulated, if not failing. With unemployment at its peak, job seekers are choosing to migrate (especially to South Africa) or to scrape by in the informal sector as alternatives. Although the Mugabe regime crafted a democratic labour market policy after the war of liberation, poor governance and authoritarian rule seem to have impeded effective implementation. In short, the Zimbabwean liberation war only resulted in the emergence of a non-racist but less-important labour market for the vast majority of Zimbabweans.

Namibia and South Africa present almost similar historical pictures from the struggle against apartheid to the emergence of neoliberal labour markets in the post-apartheid period. The post-apartheid political regime in Namibia under SWAPO seems to have effectively ended the racialised labour market through labour market reforms geared towards the democratisation of the labour market. But the labour market in Namibia continues to be characterised by notable levels of unemployment, low job security and employment inequality, because the government has instituted neoliberal economic policies.

Post-apartheid South Africa has also moved from a legally racialised labour market to a democratic labour market. This change is of course the result of a combined armed, societal and diplomatic struggle led by the ANC and the labour movement. With the change of government and ascent of the ANC to political power in 1994, the people of South Africa, especially workers and the unemployed looked forward to a general improvement in the South African labour market. Although the ANC government has instituted reforms and crafted employment policies to address labour market issues, studies suggest that the adoption of neoliberalism has worsened unemployment and contributed to outsourcing of jobs and employment inequality in the country (Baker 2010; Bateman 2019; Magubane 2004; Satgar 2020; van der Walt 2009; Williams and Taylor 2000).

Conclusion

This chapter has raised three important factors in the history of post-war labour market reconstructions in the SADC region. Firstly, the creation of wage labour markets in the region was a product of colonisation. After colonialists settled in the region, they successively created slave labour markets, forced labour markets and racialised wage labour markets. The racial injustices and inequalities in these wage labour markets induced Africans into engaging in armed struggle for liberation from settler and colonial exploitation. These struggles eventually led to political independence and subsequent changes in labour markets. The change in labour markets also applies to countries that experienced post-independence civil wars. Because of this, it can be argued that war and labour markets shape each other. As the chapter has shown, the realities of labour markets cause wars. And after wars, there are changes in labour market regimes.

Secondly, post-war political regimes of the countries in this region have rarely implemented economic and labour market policies that meet the needs of their populations. Thus liberation seems to be a myth that has turned into an illusion for most citizens of post-war politically independent countries. The end of colonialism and liberation war produced labour markets that are actually different from those that people expected before and during the wars. While democratic political regimes in the region have attempted to improve the labour markets of their countries, authoritarian regimes seem to have disarticulated the functioning of their countries and labour markets. This chapter has, however, not substantively engaged with a comparative analysis of labour market regimes in countries that descended into civil war and those that did not in the post-colonial period and proposes further research on the matter, as recommended in Chapter 8 of this book. The discussion in this chapter, however, shows that there is every reason to believe that the contribution of post-war political regimes to the reconstruction of post-war labour markets could be positive or negative.

Finally, the discussion in this chapter has shown how the dynamics of the international system simultaneously influence wars and processes of post-war labour market reconstruction. The Cold War in particular shaped both wars and processes of post-war labour market reconstruction in the SADC region before 1990. Similarly, the trend towards market economics and neoliberalism is explained by the current international dynamics that followed the end of the Cold War. Needless to say, the dynamics of the international system have a huge impact on post-war labour market reconstructions.

References

Alden, Chris. 2007. "Human Security and the Limits of International Intervention: A Comparison of Post-Conflict Angola and Mozambique." Pp. 142–58 in *Protecting Human Security in a Post 9/11 World*. Springer.

Alexopoulou, Kleoniki. 2018. "An Anatomy of Colonial States and Fiscal Regimes in Portuguese Africa: Long-Term Transformations in Angola and Mozambique, 1850s–1970s." PhD, Wageningen University and Research, Wageningen.

Arkin, Marcus. 1960. "Strikes, Boycotts-and the History of Their Impact on South Africa." *South African Journal of Economics* 28(4):303–18.

Arrighi, Giovanni, Nicole Aschoff and Ben Scully. 2010. "Accumulation by Dispossession and Its Limits: The Southern Africa Paradigm Revisited." *Studies in Comparative International Development* 45(4):410–38.

Baker, Peter A. 2010. "From Apartheid to Neoliberalism: Health Equity in Post-Apartheid South Africa." *International Journal of Health Services* 40(1):79–95.

Ball, Jeremy. 2005. "Colonial Labor in Twentieth-Century Angola." *History Compass* 3(1):1–9.

Barchiesi, Franco. 2007. "Labor and Social Citizenship in Colonial and Postcolonial Modernity: South African Perspectives in a Continental Context." *Review (Fernand Braudel Center)* 30(1):17–43.

Bateman, Milford. 2019. "South Africa's Post-Apartheid Microcredit Experiment: Moving from State-Enforced to Market-Enforced Exploitation." Pp. 69–97 in *Forum for Social Economics*, Vol. 48. Taylor & Francis.

Bauer, Gretchen. 2002. "Labor and Democracy in Namibia, 1971–1996." *Sister Namibia*:38–39

Bezuidenhout, Andries and Sakhela Buhlungu. 2011. "From Compounded to Fragmented Labour: Mineworkers and the Demise of Compounds in South Africa." *Antipode* 43(2):237–63.

Birmingham, David. 1982. "A Question of Coffee: Black Enterprise in Angola." *Canadian Journal of African Studies/La Revue canadienne des études africaines* 16(2):343–46

Bond, P., D. Miller and G. Ruiters. 2001. "The Production, Reproduction and Politics of the Southern African Working Class: Regional Class Struggle During Sustained Economic Crisis." in *Socialist Register*, edited by L. Panitch. London: Merlin Press.

Bonner, Philip, Jonathan Hyslop and Lucien Van Der Walt. 2007. "Rethinking Worlds of Labour: Southern African Labour History in International Context." *African Studies* 66(2–3):137–67.

Botes, Anri. 2013. "The History of Labour Hire in Namibia: A Lesson for South Africa." *Potchefstroom Electronic Law Journal/Potchefstroomse Elektroniese Regsblad* 16(1):505–36.

Breckenridge, Keith. 2008. "Power without Knowledge: Three Nineteenth Century Colonialisms in South Africa." *Journal of Natal and Zulu History* 26(1):3–30.

Brunger, Alan G. 2003. "The Geographical Context of Planned Group Settlement in Cape Colony: The 1820s British Emigrants." *Journal of Historical Geography* 29(1):51–72.

Carpenter, Louisa. 2012. *Conflict Minerals in the Congo: Blood Minerals and Africas Under-Reported First World War*. United States: Suffolk University.

Chabal, Patrick. 2001. "Angola and Mozambique: The Weight of History." *Portuguese Studies* 17(1):216–32.

Chinguno, Crispen. 2014. "Trade Union Leadership and Gender: Cases from Zimbabwe." Pp. 180 in *Labor and Globalization. Visibility and Voice for Union Women: Country Case Studies from Global Labour University Researchers*, edited by A. O. Britwum and S. Ledwith. München: Rainer Hampp Verlag.

Cooper, Allan D. 1999. "The Institutionalization of Contract Labour in Namibia." *Journal of Southern African Studies* 25(1):121–38.

Cooper, Allan D. 2007. "Reparations for the Herero Genocide: Defining the Limits of International Litigation." *African Affairs* 106(422):113–26.

de Grassi, Aharon. 2018. "Beyond Gatekeeper Spatial Metaphors of the State in Africa: Relational Geo-Histories of Angola." *Third World Thematics: A TWQ Journal* 3(3):398–418.

Denoon, Donald J. N. 1967. "The Transvaal Labour Crisis, 1901–6." *The Journal of African History* 8(3):481–94.

Dooling, Wayne. 2005. "The Origins and Aftermath of the Cape Colony 'Shottentot Code' of 1809." *Kronos: Journal of Cape History* 31(1):50–61.

Eichstaedt, Peter. 2011. *Consuming the Congo: War and Conflict Minerals in the World's Deadliest Place*. Chicago: Lawrence Hill Books.

Feinstein, Charles Hilliard. 2005. *An Economic History of South Africa: Conquest, Discrimination, and Development*. New York: Cambridge University Press.

Fenwick, Colin. 2006. "Labour Law in Namibia: Towards an 'Indigenous Solution'?" *South African Law Journal* 123(4):665–99.

Grieves, Forest L. 1977. *Conflict and Order: An Introduction to International Relations*. Houghton Mifflin: Boston.

Grün, Carola. 2004. "Direct and Indirect Gender Discrimination in the South African Labour Market." *International Journal of Manpower* 25(3–4):321–42.

Harris, Karen. 1991. "The 1907 Strike: A Watershed in South African White Miner Trade Unionism." *African Historical Review* 23(1):32–51.

Harris, Karen Leigh. 2013. "Indentured "Coolie" Labours in South Africa: The Indian and Chinese Schemes in Comparative Perspective." *Diaspora Studies* 6(2):92–102.

Hay, Douglas and Paul Craven. 2005. *Masters, Servants, and Magistrates in Britain and the Empire, 1562–1955*. University of North Carolina Press.

Henderson, Ian. 1972. "White Populism in Southern Rhodesia." *Comparative Studies in Society and History* 14(4):387–99.

Iliffe, John. 1999. "The South African Economy, 1652–1997." *The Economic History Review* 52(1):87–103.

Johnson, David. 1992. "Settler Farmers and Coerced African Labour in Southern Rhodesia, 1936–46." *The Journal of African History* 33 (1):111–28.

Kagan-Guthrie, Zachary. 2011. "Repression and Migration: Forced Labour Exile of Mozambicans to São Tomé, 1948–1955." *Journal of Southern African Studies* 37(3):449–462.

Keese, Alexander. 2012. "The Constraints of Late Colonial Reform Policy: Forced Labour Scandals in the Portuguese Congo (Angola) and the Limits of Reform Under Authoritarian Colonial Rule, 1955–61." *Portuguese Studies* 28(2):186–200.

Keese, Alexander. 2013. "Searching for the Reluctant Hands: Obsession, Ambivalence and the Practice of Organising Involuntary Labour in Colonial Cuanza-Sul and Malange Districts, Angola, 1926–1945." *The Journal of Imperial and Commonwealth History* 41(2):238–58.

Keese, Alexander. 2015. "Developmentalist Attitudes and Old Habits: Portuguese Labour Policies, South African Rivalry, and Flight in Southern Angola, 1945–1974." *Journal of Southern African Studies* 41(2):237–53.

Kondylis, Florence. 2010. "Conflict Displacement and Labor Market Outcomes in Post-war Bosnia and Herzegovina." *Journal of Development Economics* 93(2):235–48.

Kyle, Steven C. 2014. "Angola and Mozambique: Marxist or Market?" in *Routledge Handbook of the History of Global Economic Thought*, edited by V. Barnett. London: Routledge.

Landau, Ingrid, Evance Kalula and Colin Fenwick. 2015. *Labour Law: A Southern African Perspective*. Geneva International Institute for Labour Studies.

Likuwa, Kletus and Napandulwe Shiweda. 2017. "Okaholo: Contract Labour System and Lessons for Post Colonial Namibia." *Mgbakoigba: Journal of African Studies* 6(2):26–47.

Likuwa, Kletus Muhena. 2015. "Colonialism and the Development of the Contract Labour System in Kavango." Pp. 105–126 in *Re-Viewing Resistance in Namibian History*, edited by J. Silverster. Windhoeck University of Namibia Press.

Magubane, Zine. 2004. "The Revolution Betrayed? Globalization, Neoliberalism, and the Post-Apartheid State." *The South Atlantic Quarterly* 103(4):657–71.

Malaba, Luke. 1980. "Supply, Control and Organization of African Labour in Rhodesia." *Review of African Political Economy* 7(18):7–28.

Maqutu, Lindiwe Nomachezi. 2020. "The Postcoloniality of Labour Law: A South African Perspective." Doctor of Philosophy (PhD) in Law, College of Law and Management Studies, University of KwaZulu-Natal, Durban.

Mlambo, Alois S. 2014. *A History of Zimbabwe*. Cambridge: Cambridge University Press.

Moodie, T. Dunbar and Vivienne Ndatshe. 1994. *Going for Gold: Men, Mines, and Migration*. Berkeley: Univerity of California Press.

Moorsom, Richard. 1977. "Underdevelopment, Contract Labour and Worker Consciousness in Namibia, 1915–72." *Journal of Southern African Studies* 4(1):52–87.

Ncube, Don. 1985. *The Influence of Apartheid and Capitalism on the Development of Black Trade Unions in South Africa*. Skotaville Publishers Johannesburg.

Nshimbi, Christopher Changwe, Inocent Moyo and Jussi P. Laine. 2020. *Borders, Sociocultural Encounters and Contestations*. Abingdon, Oxon: Routledge.

O'Laughlin, Bridget. 2002. "Proletarianisation, Agency and Changing Rural Livelihoods: Forced Labour and Resistance in Colonial Mozambique." *Journal of Southern African Studies* 28(3):511–30.

Olsen, Paulette J. 1995. "Prospects for Mediation and Arbitration in Southern Africa." Pp. 153–161 in *The New Labour Market: Reconstruction, Development and Equality*, edited by E. Kalula and D. Woolfrey. Cape Town, Durban, and Johannesburg: Labour Law Unit (University of Cape Town), Centre for Socio-legal Studies (University of Natal), Centre for Applied Legal Studies (University of the Witwatersrand).

Oosthuizen, Morné and Haroon Bhorat. 2005. *The Post-Apartheid South African Labour Market*. Cape Town: University of Cape Town.

Paton, Bill. 1994. *Labour Export Policy in the Development of Southern Africa*. London: Palgrave Macmillan.

Pérez Niño, Helena and Philippe Le Billon. 2014. "Foreign Aid, Resource Rents, and State Fragility in Mozambique and Angola." *The ANNALS of the American Academy of Political and Social Science* 656(1):79–96.

Quinn, Stephanie. 2020. "Infrastructure, Ethnicity, and Political Mobilization in Namibia, 1946–87." *The Journal of African History* 61(1):45–66.

Raftopoulos, Brian and Alois Mlambo. 2009. *Becoming Zimbabwe. A History from the Pre-Colonial Period to 2008: A History from the Pre-Colonial Period to 2008*. Johannesburg African Books Collective and Jacana Media.

Rennie, J. Keith. 1978. "White Farmers, Black Tenants and Landlord Legislation: Southern Rhodesia 1890–1930." *Journal of Southern African Studies* 5(1):86–98.

Samset, Ingrid. 2002. "Conflict of Interests or Interests in Conflict? Diamonds & War in the DRC." *Review of African Political Economy* 29(93–94):463–80.

Satgar, Vishwas. 2020. "The South African Precariat, Covid-19 and# Bignow." *Global Labour Journal* 11(2)

Scott, Peter. 1954. "Migrant Labor in Southern Rhodesia." *Geographical Review* 44(1):29–48.

Swiegers, Gertrude Maylene. 2014. "Britain and the Labour Question in South Africa: The Interaction of State, Capital, Labour and Colonial Power, 1867–1910." University of the Free State.

Takirambudde, Peter Nanyenya. 1995. "The New Labour Law in Southern Africa: An Overview." Pp. 132–152 in *The New Labour Market: Reconstruction, Development and Equality*, edited by E. Kalula and D. Woolfrey. Cape Town, Durban, Johannesburg: Labour Law Unit (University of Cape Town), Centre for Socio-legal Studies (University of Natal), Centre for Applied Legal Studies (University of the Witwatersrand).

Terreblanche, Sampie, Karl Von Holdt and Solomon Johannes Terreblanche. 2002. *A History of Inequality in South Africa, 1652–2002*. Pietermaritzburg and Sandton: University of Kwazulu Natal Press.

Tshimpaka, Leon Mwamba and Saint José Inaka. 2020. "Long March to South Africa: The Breaking of Colonial Borders Through Migrant Smuggling of Congolese." Pp. 89–111 in *Borders, Sociocultural Encounters and Contestations: Southern African Experiences in Global View*, edited by C. C. Nshimbi;I. Moyo and J. P. Laine. London and New York: Routledge.

Tshimpaka, Leon Mwamba and Christopher Changwe Nshimbi. 2022. "Scaffolding the State: Faith-Based Organisations and Application of Democratic Principles in the DRC and Zambia." *Religion, State & Society* 50(1):76–95.

Vahed, Goolam. 2019. "'An Evil Thing': Gandhi and Indian Indentured Labour in South Africa, 1893–1914." *South Asia: Journal of South Asian Studies* 42(4):654–74.

van der Walt, Lucien. 2007. "The First Globalisation and Transnational Labour Activism in Southern Africa: White Labourism, the IWW, and the ICU, 1904–1934." *African Studies,* 66(2–3):223–51.

van der Walt, Lucien. 2009. "Labor in South Africa: A Sleeping Giant?" *Real World Labour: A Reader in Economics, Politics & Social Policy*:27–42.

Van Onselen, Charles. 1973. "Worker Consciousness in Black Miners: Southern Rhodesia, 1900–1920." *The Journal of African History* 14(2):237–55.

Vindevogel, Sofie, Kathleen Coppens, Ilse Derluyn, Maarten De Schryver, Gerrit Loots and Eric Broekaert. 2011. "Forced Conscription of Children During Armed Conflict: Experiences of Former Child Soldiers in Northern Uganda." *Child Abuse & Neglect* 35(7):551–62.

Vos, Jelmer. 2014. "Work in Times of Slavery, Colonialism, and Civil War: Labor Relations in Angola from 1800 to 2000." *History in Africa* 41:363–85.

Walter, Ralston McBareth. 1973. "Racism and Revolution: A Case Study of Angola." Arts and Social Sciences: Department of Political Science, Sociology and ....

Webster, Edward, Rob Lambert and Andries Beziudenhout. 2008. *Grounding Globalization: Labour in the Age of Insecurity*. Malden, MA: Wiley.

Werner, Wolfgang. 1993. "A Brief History of Land Disposession in Namibia." *Journal of Southern African Studies* 19(1):135–46.

Willcox, Walter F. 1931. *International Migrations. Vol. II*. New York: National Bureau of Economic Research.

Williams, Paul and Ian Taylor. 2000. "Neoliberalism and the Political Economy of the 'New' South Africa." *New Political Economy* 5(1):21–40.

Wood, Geoffrey and Mark Harcourt. 1998. "The Rise of South African Trade Unions." *Labour Studies Journal* 23(1):74–92.

Worden, Nigel. 2011. *The Making of Modern South Africa: Conquest, Apartheid, Democracy*. Malden, MA: Wiley.

CHAPTER 3

Conceptual and Theoretical Issues in the Making, Unmaking and Remaking of Labour Markets in the SADC Region

This chapter engages with definitional and theoretical issues on labour markets in relation to the region that houses the Southern African Development Community (SADC). The chapter provides the lenses through which can be analysed the creation, break down and reconstruction of labour markets in a region such as the SADC and member states that have been exposed to war and then go on to become post-war states. Considering the focus of the book, and that to the best of our knowledge no study of this sort that focuses on the SADC region exists, the chapter first explores the basic concepts of labour market; labour market institutions; the nation-state wherein such institutions exist, break down and are reconstituted; power; and the international system. At the theoretical level, the chapter engages with classical, neoclassical and institutionalist labour market paradigms; labour segmentation theory; and Peck's theory of social regulation. In our assessment, combining these perspectives optimises the possibility of most aptly explaining the post-war labour market reconstruction particularly in member states of the SADC. Furthermore, deploying Peck's theory of social regulation has particular relevance to the SADC member country on which the book zooms in—the DRC—to engage in detailed analysis of the complexities in the dynamic making, unmaking and remaking of post-war labour markets. In this chapter, we demonstrate that the links between the past and the present within the

S. J. Inaka et al., *The Reconstruction of Post-War Labour Markets in The Southern African Development Community*,
https://doi.org/10.1007/978-3-031-37357-2_3

Congolese labour market are consistent with the argument of Peck that labour markets are historically constructed and embedded. Hence, we centrally draw and attempt to expand on Peck (1996) and the theory of global labour history in this chapter. This is in order to enable us to examine the making, unmaking and remaking of the labour market in the SADC region and particularly the DRC in subsequent chapters of the book.

We begin by looking at basic concepts including, the labour market, labour market institutions, the state, power and the international system. Thereafter, we examine some relevant theoretical approaches to post-war labour market reconstruction. These include the classical, neoclassical and institutionalist theories; labour market segmentation theory; Peck's theory of social regulation; and global labour history.

Conceptual Issues

We make use of five concepts in this book, to help examine the construction, unmaking and reconstruction of labour markets in the SADC region and in the DRC. The notions provide conceptual tools with which to analyse the said labour market dynamics in the region and the country. The concepts are labour market, labour market institutions, the state, power and the international system. Aside from conceptual framing, they also constitute the pragmatic context(s) within which the making, unmaking and remaking of labour markets in the region and country occurs in reality. We briefly examine each of them in turn in the following subsections, highlighting debates around them and elucidating ambiguities. We start with the concept that is basic to the book, the labour market.

The Labour Market

The concept of the labour market is heavily debated between neoclassical economists and other social scientists. Neoclassical economists understand the labour market as an imaginary market in which employers buy labour and employees sell labour as a commodity under contractual terms (Peck 1996). Scholars who view the labour market in this way consider both employers and employees to enjoy the freedom of choice in taking up and offering labour, respectively (Wilthagen and Rogowski 2002). For them, the labour market is a competitive space between the buyers of labour (employers) and the sellers of labour (employees) and is marked by competition over jobs between the sellers of labour (Kaufman 2007). This competition leads to an equilibrium of price that mirrors the balance between supply and demand (Fevre 1992).

Kalleberg and Sorensen (1979) summarise the meaning of the notion of labour market as broadly referring to

> the institutions and practices that govern the purchase, sale, and pricing of labour services. These structures include how workers are distributed among jobs and the rules that govern employment, mobility, the acquisition of skills and training, and the distribution of wages and other rewards obtained contingent upon participation in the economic. (Kalleberg and Sorensen 1979: 351–52)

This conceptualisation of the labour market is useful for analysing labour market reconstruction in post-war settings because it offers a wide range of indicators including, labour market institutions; labour market behaviours; labour demand; labour supply; labour allocation; career advancement; labour market legislation and regulation; wages; and fringe benefits.

In contrast, institutional economists and some sociologists (see i.e. Braverman 1974; Streeck 2005, 2010) argue that labour is neither a commodity nor a homogeneous good. As workers are human beings, it is impossible to detach labour from persons who sell or possess it. Moreover, labour and workers are embedded in society, so that other social, political, historical and institutional factors influence the choices that individuals make in labour markets.

There are also other scholars who assert that free and unregulated labour markets are neither free nor sufficiently fair (e.g. Peck 1996; Fine 2003). Peck (1996), for instance, approaches the labour market as a structure that is causally conjectural. According to him, national, regional and global tendencies articulate particular labour market environments. For Peck, labour markets are rather distinctively flexible on the basis of the socio-political contexts in which contradictions in labour agglomerations are resolved. Hence, You and Chang (1993), who equally find the concept of free labour market problematic, present empirical evidence from Korea that challenges the conventional wisdom that holds economic models of equity and growth to be responsible for growth based on free labour markets. They show that the state in Korea assumes an active role in managing the labour market and thereby heavily regulates it.

Labour Market Institutions

Labour market institutions broadly constitute public and private organisations that regulate the workplace. They include, among others, collective bargaining organisations, trade unions or trade/craft associations, employers'/business associations, and training and skills-transfer organisations. Labour market institutions also include entities that 'redistribute income, including pensions, income support for the unemployed and the poor, as well as public social services' (Berg 2015: 1). They have a role to play at several points in the labour market, including the demand side, the supply side, regulation and legislation. For that reason, Peck (1996: 100–101) argues that it is critical to examine these institutions in detail. This analysis should take into consideration local and international realities as these institutions are embedded in local, national and international contexts (Peck 1996: 101).

The idea that protective labour market institutions impede the functioning of the labour markets and cause the increase in unemployment is central to neoclassical economics theory (Freeman 2010). This theory considers the institutions to stifle the capacity 'of private parties to freely set quantities and prices' (Saint-Paul 2000: 1). The theory also views the efforts of labour market institutions to promote minimum wages, unemployment benefits and protection as rigid and harmful to job creation and that it leads to increases in unemployment (Saint-Paul 2000).

Accordingly, neoclassical economists advocate for the liberalisation of labour markets based on the assumption that it boosts investments and promotes high employment. To do so, they suggest that countries institutionalise free trade, reduce the role of the state in labour markets, weaken the power of trade unions, and privatise public companies. In short, they argue that the deregulation of the labour market would lead to increases in investment and that would in turn lead to economic growth and high employment (Berg and Kucera 2008: 12). Ultimately, they call for a reduction in the role of the state.

It is important to take the neoclassical economics theory into consideration here as it influentially underpins the views and approach of international financial institutions such as the International Monetary

Fund (IMF) and the World Bank, which have great weight in post-war labour market reconstruction processes around the world, the DRC included. The influence of this theory on these institutions is such that they consider employment protection legislation, because it is 'generous' to unemployment, as likely to lead to high unemployment (Howell 2004). Moreover, the World Bank and the IMF seem to be of the view that hiring staff on permanent contracts, paying large amounts in severance packages on dismissal and protecting labour rights hamper investment and job growth (Berg and Kucera 2008: 1). Similarly, they are concerned that labour market institutions undermine structural adjustment programmes (SAPs) designed to cure balance of payment deficits (Freeman 2010), which requires a shift of resources from the public to the private sector.

Criticism of neoclassical economics theory includes the observation that it tends to overstate the negative effects of labour market institutions on the functioning of labour markets. For critics who hold this view, labour market laws and labour market policies are necessary for ensuring social justice for workers in developing countries. Institutions that regulate the labour market are important in these countries because many workers there enjoy little or no protection under the law. They are excluded from social security and are often underpaid. Unlike the World Bank and the IMF, the International Labour Organization (ILO) supports the setting up and promotion of labour standards. Thus, the ILO requires member states to guarantee effective protection for workers through national labour laws, efficient labour market administration, and labour reform policies (Berg and Kucera 2008).

Many post-war countries pursue labour policies to improve and strengthen labour market institutions, despite the IMF and World Bank's neoliberal agenda. Against this background, an Institutionalist understanding of the role and importance of these institutions is critical to this study.

The State

We made the point in the previous chapter, with regard to 'methodological nationalism' in the analysis of labour history in Southern Africa, to avoid relying on the nation-state as a unit for examining labour formations. Yet, the state in the region apparently continues to be in charge of policy formulation and, led by politicians, acts in ways that suggest that it assumes responsibility over labour markets within its territorial jurisdiction. Moreover, the state, as defined by the geographical territorial area over which is exercised control, jurisdiction or sovereignty, constitutes the physical space within which labour markets are made, unmade and remade.

As a concept, the state has multiple connotations and dimensions. In this book, we draw on the definition provided by the 1933 Montevideo Convention on the Rights and Duties of States. This convention establishes benchmarks by which member states of the United Nations recognise other states in the international system. The convention sets up four criteria that have to be fulfilled for a state to be recognised in accordance with customary international law in stating that,

> [t]he state as a person of international law should possess the following qualifications: 1) a permanent population …; 2) a defined territory ..; 3) a government … and; 4) [the] capacity to enter into relations with the other states. (United Nations Treaty Collection 2013)

It is necessary to understand the difference between state and government. In liberal democracies, the government is composed of three separate bodies: the legislature or law-making body, the judiciary that oversees interpretations and applications of the law, and the executive, responsible for implementing the law and managing a country (Duverger 1988: 23). The core roles of a state are to defend its integrity and independence; to protect its citizens; to provide for its conservation and prosperity; to deliver the political advantages of statehood; and to administer the country (Duverger 1988; Mpongo-Bokako 2001).

Government represents the means through which the state fulfils these roles. The government encompasses the people and agencies designated to complete a state's purposes (Wood and Brewster 2007), including those required for the reconstruction and regulation of a post-war labour market.

Many polities in Africa do not reflect the ideal type of state as captured in these definitions. Various scholars have suggested ways in which to approach and understand African states. Many scholars, for instance, argue that patrimonialism—the tendency to ascribe the right to rule to individual rulers rather than to a ruling administration (Thomson 2010)—is a characteristic feature of African states. Rulers in patrimonial states misappropriate state resources and distribute them in ways that ensure the loyalty of their supporters. Real power in the patrimonial states is exercised through individual relationships rather than political institutions or established bureaucracies. The division between private and public is blurred, the rule of law is often undermined and even reforms are tailored to suit the personal interests of rulers (Clapham 1985). Bratton and Van de Walle (1994) describe neopatrimonial African regimes as polities ruled by chief executives who maintain authority through personal patronage rather than through ideology or law. In such states, bureaucrats perform less public service, seeking instead to accrue personal wealth and status. Neopatrimonial regimes are grounded in exchanges of favours between patrons and clients: patrons give public sector job positions and/or socioeconomic favours (licences, contracts, projects) to their clients while clients mobilise political, social and administrative support for their patrons.

Chabal and Daloz (1999) used the notion of the 'instrumentalization of disorder' to describe neo/patrimonial states. They suggest that African rulers take advantage of weak institutions in their countries to satisfy their various needs. Bayart et al. (1999) suggest that in these settings, African state bureaucracies often turn to corruption instead of performing their accredited roles.

Mbembe (2000) proposes the concepts of *commandement* and indirect private government to understand states in Africa. Strong individuals use the *commandement*, the violence of the state, as the strategy of oppression and repression to ensure submission to their authority. They also distribute gifts or donations, allocate jobs to their cronies and guarantee various favours to individuals in order to establish and perpetuate their power. An indirect private government, in turn, is a regime of power that operates to the advantage of strong individuals, who are competing mercilessly for the concentration of power and the hoarding of public assets. Although public administrations tend to resist, their fate is often to be dismembered: they become empty shells, their orders are ignored, their hierarchies are rejected and their positions abused as tradeable commodities.

Other scholars attempt to differentiate between weak, failing, failed and collapsed states and particularly those in Africa (Kreijen 2004; Rotberg 2004, 2010; Giorgetti 2010). Rotberg (2004) proposes as criteria for this differentiation the level of a state's capacity to provide a set of 'political goods' namely, human security, prevalence of the rule of law, political freedom, civil and human rights, medical and health care, education and physical infrastructures. However, some scholars express the critical view that labelling African states as failed, weak or collapsed is problematic. They argue that the labelling represents agendas, especially of Western donor countries and international financial institutions, to use the said shortcomings as pretexts for intervening and pursuing certain kinds of policies in the so-called failed states (Zoellick 2008; Grimm et al. 2014). The creators of this narrative argue that state fragility hinders peace and development and creates conducive conditions for organised crime, transnational terrorism and the proliferation of weapons (Zoellick 2008; Boege et al. 2009). On the other hand, they argue, the fragility and failure make it necessary to design development policies and provide development assistance in order to establish state institutions that are functional and effective as a prerequisite for development (Zoellick 2008). But Boege et al. (2009), for example, argue that the perception of state fragility as an obstacle to peace and development, which is so prevalent in Western political thinking, is misleading. Therefore, they propose alternative assessments and explanations of statehood in places like Africa. According to them, a wider perspective and changing the focus can help achieve a deeper theoretical and political understanding of issues at hand in the so-called fragile states.

Power

The concept of power is fundamental in the social sciences (Clegg and Haugaard 2009) and particularly in political science and political sociology (Pels 2002). However, it is a polysemous concept, triggering bottomless debates (Hermet et al. 1998). In this section, we review some of these debates and then explain why we consider Foucault's conception as particularly suitable to helping explain how power operates in post-war labour market reconstruction.

Weber (1968: 212) defines power as 'the chance of a man or a number of men to realise their own will in a command action even against the resistance of others who are participating in the action'. Alongside this notion of power, he identifies three types of authority: traditional authority, as power legitimised through respect for long-established cultural patterns; charismatic authority, drawn from the devotion felt towards a leader by subordinates who believe the leader possesses exceptional qualities; and rational-legal authority, legitimated through the enactment of rules and regulations and instituted through formal organisations. This conception of power has been discussed and reformulated by several thinkers. Dahl (1961) argues that power has to be analysed in the context of organisations and structures, and particularly in the roles of ruling elites. He believes that certain elites exercise power in society and compel those who are subjected to them to follow their private preferences (Dahl 1971). Bachrach and Baratz (1962) argue that there are two faces of power: the overt face, linked to the way decisions are made, and the covert face, the capacity to resist a decision. Lukes (2005), in contrast, challenges Weber's conceptualisation, arguing that it overlooks the latent dimension of power, namely the relation between political preference and *real interests*. In his view, the analysis of power ought to look at 'the entire political agenda, in order to examine its adequacy to the true interest of various groups' (Lukes, as cited in Sadan 2004: 37).

Foucault (1975) argues that power is not a property or commodity that can be owned and/or exerted by institutions, persons, or groups of individuals. Rather, he conceptualises power as omnipresent and multidirectional, something that 'operates from the top-down and bottom-up' and that 'occurs at sites of all kinds and sizes, including in the most minute and most intimate circumstances, such as the human body' (Foucault 1975: 32). For him, power relations are mobile, asymmetrical, unbalanced and dependent on culture, place and time; and they involve discourse (systems of ideas), which allows the possibility of resistance (Foucault 1975: 119). Power gives the conditions in which people's conduct can be corrected, regulated, checked and controlled through surveillance (Sadan 2004: 58–59). Foucault rejects the widespread notion of sovereignty, however, namely the conception that independent states are free from outside power (Foucault, as cited in Dean 2010: 330). His rejection is premised on the reason that outsiders will always interfere in the internal affairs of other countries.

Foucault (1975; see also Kelly 2010: 37–38) distinguishes six characteristics of power: (a) Power is impersonal, or subjectless: it is not guided by the will of individual subjects (Foucault 1975: 27); (b) Power is relational: it always exists as a relation between people; it is not a quantum possessed by people (Foucault 1975: 27); (c) Power is decentred: it is not concentrated in a single individual or class (Foucault 1975: 27); (d) Power is multidirectional: it flows not only downwards, from the more to the less powerful, but also upwards, even if it is nevertheless non-egalitarian; (e) Power is strategic: it has a dynamic of its own and is intentional (Foucault

1975: 26) and; (f) Power is intertwined with knowledge: knowledge gives birth to technologies or practices that exert power (Foucault 1975: 21).

Some scholars decry the Foucauldian conception of power, however, on the basis that it lacks an explicit basis in normative assumptions (Kelly 2010: 5). Such scholars argue that Foucault does not provide a general theory of power. Other scholars (Kelly 2010: 4), argue on the contrary; that Foucault's conception of power is consistent and coherent since it approaches power from its manifestation at the social micro-level.

For Foucault, power and resistance cannot be decoupled. Human beings always seek ways to escape from bonds of power, even if this takes the form of an infinite diffusion of resistance and micro-revolts. Other Foucauldians (Dean 2010; Kelly 2010) suggest that Foucault's concept of governmentality clearly addresses the adequacy between a government's policies and its citizens' desiderata, and processes of making societies governable. Governmentality, in its wider sense, also refers to the governance of a given state, the set of mental, technical and cultural practices that influence the way in which people are governed (Dimier 2010).

Foucault's understanding of power as omnipresent and multidirectional, in a web of mobile, asymmetrical and unbalanced power relations, is very useful to approach and understand the complex roles played by multiple actors in post-war labour market reconstruction processes such as in the DRC. We draw on Foucault's notions of sovereignty and governmentality in this book to be more sensitive to how the international system and local realities interact and influence these reconstructions.

The International System

The international system is an assemblage of relationships among states (the main actors) and non-state actors (like international organisations, non-governmental organizations (NGOs), multinational corporations (MNCs), or transnational criminal networks) that is structured according to certain rules and patterns of interaction (Hill 2015). The activities of these actors are grounded in a distribution of power relations that shapes the international system at any given period (Hyde-Price 2007). This is because these power relations create a system of polarities that harden into blocs that exert power on an international level (Ikenberry et al. 2009). These polarities evolve with changes in the stages of international capitalism (Dugin 2018). The polarities also influence labour markets as well as the outcomes of post-war labour reconstruction. There are three types of 'polarity systems': the multipolar system, the bipolar system and the post-Cold War era (Hyde-Price 2007).

The Multipolar System

In a multipolar system, multiple states possess a considerable degree of military, political, cultural and economic influence in the world. This implies that the distribution of power among states is concentrated at several poles of power (Scott 2013). Multipolarity characterised the Westphalian political system that generally defined the world from 1648 to 1945 (Hyde-Price 2007).

The multipolar system co-evolved with two successive stages of Western capitalism: mercantilism (from the sixteenth to the early nineteenth century) and competitive capitalism (from the late nineteenth century to the 1940s). Mercantilism was at the core of the triangular trade of slaves from Africa (Williams 2014). It allowed European merchant capitalists, who owned the means of buying, transporting and selling, to play a dominant role in organising relations of production. Competitive capitalism occurred during the early stage of colonisation, before WWI (Albritton 1991). As seen in the previous chapter, many colonisers enacted laws and set up labour policies aimed at installing and reinforcing the capitalist labour market, and retaining a steady and available supply of cheap labour by force (Sender and Smith 2003).

The Bipolar System

In a bipolar international system, two competing superpowers, associated with their respective allies, dominate the world. Bipolarity is exemplified by what happened between 1947 and 1990. During this period the USA and the Soviet Union represented two opposite poles of power, supported by their respective allies (Hyde-Price 2007; Scott 2013).

This period coincided with capitalism characterised by the internationalisation of Fordism and Keynesianism, after WWII (Albritton 1991). The latter facilitated stable economic growth and living standards in many advanced capitalist countries from 1949 to 1973 (Harvey 1989). The era was additionally marked by the emergence of stable rule-governed institutions that regulated labour markets in Western countries, guaranteed security of employment and strengthened trade unions (Gartman 1998).

The post-WWII era was also a time of transition to independence from colonialism for many African countries. Many colonial authorities tried to improve labour regimes through the amelioration of labour conditions, the partial authorisation of trade unionism, the implementation of new labour acts and by increasing job opportunities for African workers (Freund 1988; Buhlungu 2010; Britwum and Dakhli 2019; Freund 2019).

By the end of the 1960s, advanced capitalist countries experienced a decline in wages and levels of profitability and productivity, accompanied by an increase in unemployment and inflation. These factors led to a crisis in the Fordist approach, marked particularly by the collapse of the Bretton Woods system in 1971 and the oil crisis of 1973 (Harvey 2007).

In mitigation of this crisis, the USA and the UK began to implement a neoliberal approach to their economies in the early 1980s (Standing 2011: 5). Neoliberalism expanded to other capitalist countries in the 1990s (Gartman 1998: 133). With the end of the Cold War in 1989 and that of the bipolar world order, neoliberal economic and political principles became the dominant model for labour markets in most post-war countries (Inaka 2020).

The Post-Cold War Era

'On 3 December 1989, the leaders of the USA and the USSR declared an end to the Cold War, after two days of talks at the Malta summit' (Spohr and Reynolds 2016: 1). This led to a shift in the international system from bipolarity to a new globalised system, termed by some scholars as the post-Cold War era (Spohr and Reynolds 2016).

There is, however, an ongoing debate whether this globalised system tends more towards a unipolar system or more towards a multipolar one. Some scholars posit that the USA became the sole superpower and established its political hegemony in the new world order (Hansen 2010; Thalakada 2012). Thalakada (2012) demonstrates how the USA and its allies persuaded the rest of the world to adhere to neoliberal political and economic ideologies and practices. This is countered by Scott (2013: 43), who argues that there is concentration of power in several poles or multiple centres of power.

The rise of neoliberalism in the post-Cold War era led to privatisation, liberalisation and deregulation of labour markets (Harvey 2007).

Theoretical Issues

Most labour market theories are grounded in classical, neoclassical and institutionalist paradigms. We begin our discussion of theories that relate to post-war labour market reconstruction in this section with these paradigms. We then go on to examine two theories—labour market segmentation theory and the theory of social regulation—because they have greater capacity to explain the reality on the ground regarding the making, unmaking and remaking of labour the market in the DRC.

Classical, Neoclassical and Institutionalist Labour Market Paradigms

At the core of the classical labour market paradigm is the idea that the labour market is analogous to the capital and commodities market. The labour market is, thus, a regulating factor on prices, salaries and the maintenance of balance between the supply and demand for labour (Kaufman 2007). The paradigm considers full employment as the normal situation in the labour market, and unemployment as an abnormality (Keynes [1935] 2018: 19–20). Full employment is given when the demand and supply of labour are equal. Unemployment, in turn, is the product of a rigid wage structure (Pigou, cited in Aldrich 1979: 321–22).

The classical labour market paradigm has three limitations. First, while full employment is an ideal situation in a capitalist economy, the underlying reality is a high rate of unemployment (Keynes [1935] 2018: 20). Second, the assumptions of perfect competition, perfect information and homogeneity of labour are unrealistic for an analysis of labour markets. Third, classical theory fails to consider labour as a human factor of production (Peck 1996). All three limitations apply to the Congolese situation, for which reason the classical labour market paradigm is not sufficiently helpful to the discussion in this book.

The neoclassical economics paradigm attempts to address the limitations of the classical theory by further suggesting that, the labour market process is driven by employers, workers and their unions (Yoder and Heneman 1959; Brožová 2015). The neoclassical economics paradigm considers labour to be an immaterial commodity of the capitalist economy that can be bought and sold on the market, with prices determined by supply and demand (Streeck 2005). Workers are thus paid proportionately to their productivity, and price relations drive labour market (Fine 2003).

Yet, as pointed out by the institutionalist scholars, labour is fictitious (Kaufman 2007). Polanyi (1944) shows how labour cannot be produced and sold in the market like other commodities. Granovetter (1985) and Peck (1996) are in accord with Polanyi in demonstrating how the labour market is a socially embedded institutional phenomenon; individual workers cannot be removed from the social relations in which they are embedded. For these reasons, the neoclassical economics paradigm does not allow us to holistically explain the course of post-war labour market reconstruction.

The institutionalist paradigm of labour markets accommodates the concerns raised against the classical and neoclassical theories. It approaches the labour market as a phenomenon grounded in social norms. Thus, collective bargaining, labour rules and practices, skills divisions and educational systems are drawn on to explain how and why labour market structures are resistant to rapid change (Wootton, as cited in Rowe 2009). Critically, institutionalists understand labour markets to embody values that extend outside the commodity market or exchange nexus that cannot be reduced to crude notions of supply and demand (Rubery 2002).

In the institutionalist paradigm, the reconstruction of institutions that provide political stability and efficient administration needs to precede political and economic reforms in post-war settings (Paris 2004; Cramer 2008). This is because a weakness in institutions charged with regulating political and economic competition can lead to the failure of reforms, and even to the re-emergence of armed conflict.

The institutionalist perspective is important in a setting such as post-war DRC, to explain the functioning of labour markets and describe institutions that might be characterised as weak, failed, or collapsed. The institutionalist paradigm is also the foundation on which the following two theories relevant to the discussion in this book are based.

Labour Market Segmentation Theory

Three approaches to labour market segmentation theory exist: the dualist, radical and the Cambridge School (Peck 1996). We discuss each in turn.

Segmentation theory began with Doeringer and Piore (1971) notion of the 'dual labour market'. The notion implied that labour markets are divided into two parts: a primary (internal, core) and a secondary (external, periphery) market. Doeringer and Piore (1971: 45) describe the two in the following way:

> Jobs in the primary market possess several of the following characteristics: high wages, good working conditions, employment stability, chances of advancement, and equity and due process in the administration of work rules. Jobs in the secondary market, in contrast, tend to have low wages and fringe benefits, poor working conditions, high labor turnover, little chance of advancement, and often arbitrary and capricious supervision.

Doeringer and Piore emphasise technological developments and industrial structures that create internal labour markets and require on-the-job training or firm-specific skilled employees in the primary sector. This theory challenges the notion of human capital, which postulates that a relationship exists between educational achievement and workers' rewards, and that the educational system serves to prepare workers and screen them for their destined roles under capitalist employment (Doeringer and Piore 1971).

The dualist model also challenges the classical economics assumption of homogeneity in the labour market. Dualist theorists explain the state of labour markets through the existing qualitative differences within jobs and job structures (Wachter et al. 1974; Barbieri and Cutuli 2016). They critique the neoclassical notion of human capital by demonstrating that employers' discriminatory strategies shape the demand for labour (Peck 1996). Dualists also suggest that labour markets are socially and institutionally constructed, hence bringing the neoclassical assumption of the self-regulating labour market into question (Kalleberg and Sorensen 1979; Marx 2012; Barbieri and Cutuli 2016).

Radicals constitute Marxist economists, who argue that dual labour markets are founded on a capitalist labour control strategy and not on employees' skills. Their argument is informed by the history of labour market segmentation. They point to the 1919–1920 strikes and labour movements in the USA when 'employers actively and consciously fostered labor market segmentation in order to 'divide and conquer' the labor force'. This strategy allowed employers to break up the labour market in four ways: a segmentation between primary and secondary sectors; a segmentation between subordinate and independent employees (in the primary sector); a segmentation by sex; and a segmentation by race (see, for example Reich et al. 1973: 360–64).

Proponents of neoclassical economics theory (like Wachter et al. 1974; Cain 1976) have critiqued the dualist model for being too descriptive of job types, lacking methodological and empirical credibility and therefore not offering a credible alternative to the 'powerful' neoclassical economics theory of labour. They argue that dualists overgeneralise the bipolarity between primary and secondary labour markets. Cain (1976), for example, refers to evidence of a high range of mobility of employees between these sectors, though this is countered by the neoclassicist Evans Jr (1973: 33) who suggests that such generalisation might be useful in explaining labour markets in developing countries. Notwithstanding, neoclassical theorists continue to challenge dualism on the basis of its limited application and overlooking, for example, supply-side factors and the role of the state in labour markets. Craig et al. (1985) observe that dualism tends to focus more on labour demand while overlooking supply-side factors and the roles of the state in the labour market.

The Cambridge School debates processes of social reproduction and the roles of social institutions (see Rubery and Wilkinson 1981; Craig et al. 1985). At the core of its argument is the claim that social forces, social institutions and the state shape interactions on both the demand and supply sides of labour markets. This group insists on the interaction between supply and demand, and the ways in which various sectors within labour markets overlap (Craig et al. 1985). The Cambridge School also emphasises state policies of labour legislation, welfare and education/training systems in regulating labour markets (Peck 1996; Fine 2003).

Market segmentation theory is criticised in particular on the basis of its theoretical limitations. Peck (1996) and Fine (2003) argue that it omits interactions between labour markets and the spatial context. The state as employer and political force that influences the labour market is also absent. Another weakness is the neglect of the role of the global economy and international economic relations.

Despite these challenges, segmentation theory is a useful tool for examining dual labour markets and policies that discriminated against workers on the basis of race or gender during the colonial period, as discussed in Chapter two, for example. The usefulness of segmentation theory may, however, be compromised by its failure to account for a post-war labour market that is skewed towards labour market flexibility. It is here that Peck's theory of social regulation discussed in the next section becomes important.

Peck's Theory of Social Regulation

Given the lack of a theory that can most appropriately explain the post-war labour markets of, especially, Southern Africa, we draw and expand on Peck's (1996) theory of social regulation of the labour market for our purposes. This is because labour market phenomena in post-war environments such as Southern Africa's should be examined through the effects of political, economic and social factors operating before, during and after conflicts (Krishnamurty 2007). To that end, the assumptions and precepts of Peck's theory help us achieve a more holistic understanding of post-war labour market reconstruction in Southern Africa and the DRC in particular. We attempt to demonstrate the usefulness of his theory here.

Peck argues that labour markets are constructed and work in locally specific ways. Thus, existing spatial, occupational, political, economic, and sociocultural structures have to be taken into account when wishing to regulate labour markets. By looking at how systems are rooted in the local, Peck's theory provides a framework for examining the governance of labour markets in the context of contradictory institutions (such as international institutions vs. national ones, companies vs. trade unions, etc.) and the constrictions that act on social actors.

In Peck's understanding, labour supply and labour demand are linked by processes of incorporation, allocation, control and reproduction. A range of social factors impact the demand for labour. These may include education and social and economic decisions. Without taking these factors into consideration, it is difficult to incorporate the labour market and the market cannot be regulated by simply matching supply with demand. These social processes have to be considered when shaping any regulation of the labour market. Peck points to four elements that can be used to identify these social processes: labour incorporation, labour allocation, labour control, and labour reproduction.

Labour Incorporation

'[I]ndividuals become wage earners or self-employed in the labour market' through labour incorporation (Peck 1996: 26). Labour incorporation determines the choices individuals make in the labour market and their potential to take up certain jobs (Bezuidenhout et al. 2008). For Peck, the flow of workers into the market is not simply a function of the number of available jobs. Rather, he argues that this flow depends on the intersection of various social, demographic, educational and economic factors, on institutional labour legislation and on 'autonomous social forces such as state policies, ideological norms, and family structures' (Peck 1996: 27). These social realities thus point to the fact that 'labour supply is not simply regulated by market forces' (Peck 1996: 27). In addition, studies must take into account extra-market actors and/or factors that unpredictably affect state regulation of the incorporation of labour. Overall, Peck's theory touches on conceptual elements that this book can use as lenses for understanding workers' integration in labour markets.

Labour market incorporation occurs in two main steps. The first consists of an individual's decision to look for a job. This decision can be free or unfree, with the latter, otherwise referred to as 'adverse labour incorporation', occurring in situations such as forced labour, slave labour, child labour and human trafficking (Phillips 2013). The second step consists of looking for information on job opportunities. For this, several studies employ Granovetter's (1973: 1367) theory of the strength of weak ties: jobseekers finding information on job vacancies through persons who they are not close to, such as acquaintances, friends of friends, former college friends, former workmates, or employers. The strength of these ties is 'a combination of the amount of time, the emotional intensity, the intimacy and the reciprocal services which characterize the tie' (Granovetter 1973: 1361). Compared to close friends or family members, these acquaintances are more likely to move in circles that are different from those of the job seekers and thus have access to job information the job seekers cannot access by themselves.

A large body of research has argued that context matters when it comes to strong or weak social ties. Langlois (1977), for example, argues that executives and professionals frequently use weak ties, while individuals of low status often draw on strong ties. Lin et al. (1981) underline the effectiveness of using weak ties in the free and complex labour markets of the USA and Europe. Brown and Konrad (2001) note that people tend to use weak ties in growth sectors of the economy, whereas they make use of strong ties in shrinking industries.

Granovetter's concept of strong and weak social ties is critical for the analysis of how Congolese engage in the job-search process. Phillips' notion of adverse labour incorporation is particularly relevant to situations of forced labour, including child labour, in the DRC.

Labour Allocation
The second phase of the labour market process is labour allocation, which links the demand and supply sides of the labour market by matching workers and jobs. Peck argues that an individual's achieved and/or ascribed status, as well as the social ties held, play a huge role in the person's allocation in the labour market. An individual's achieved status refers to the manner in which their human capital (qualifications, work experience, and skills) determines how he/she may be slotted within the labour market. The allocation of labour is often influenced by ascribed characteristics such as ethnicity, race, sexual orientation, gender and citizenship (Hudson 2007). In addition, 'the complex intermeshing of employers' recruitment networks and worker supply networks plays a significant role in labour allocation (Peck 1996). Peck, therefore, insists that labour allocation is based more on relations of trust than on human capital (Peck 1996: 33–34). This point is critical in analysing the ways in which employers on the DRC labour market hire Congolese employees.

Labour Control
Labour control enables the achievement of 'the interrelated processes of, first, securing appropriate labour supply; second, maintaining control within labour process; and third, effectively reproducing this set of relations' (Peck 1992: 337). It constitutes the terrain where conflict of interests between employers and employees take place. While employers seek to control employees in order to increase their productivity and the functioning of the enterprise, employees seek to defend their autonomy, the physical integrity of their labour power and their skills. No universal rules of labour control exist, but the rules rather vary historically, geographically and from one industry to another (Peck 1992). Other factors can also impact labour control, such as production, technology, plant size, the nature of the labour supply, and the system of labour regulation (Peck 1996; Bassanini and Ernst 2002; Hasan and Jandoc 2013; Almeida and Poole 2017).

Strategies of labour control are strongly influenced by local politics (Peck 1992). The role of the state in the reproduction of labour, the relationship between internal and external labour markets, and the dynamics of political movements are among factors that shape labour control (see, e.g., Cai and Pandey 2013; Haltiwanger et al. 2014; Alesina et al. 2018).

Drawing on Burawoy's (1979) typology of labour control, which distinguishes the despotic labour control of early capitalism from contemporary hegemonic labour control, Peck differentiates consent from control. Despotic labour control, in his presentation, is managerial despotism. In such a management system, workers do not have a voice in the labour process. Instead they are coerced into following instructions. Consent, in contrast, draws willingness and cooperation from workers:

> [A]chieving and maintaining a balance between consent and control in the workplace is not simply a matter of setting appropriate wage rates but involves continuous negotiation with both labour process (for example, around the introduction of new technologies) and the labour market (for example, around hiring and promotion decisions). (Peck 1996: 32–33)

Research based on Peck's approach shows the existence of various labour control strategies. Direct political control can be as blatant as deploying state agents in companies in order to monitor the activities of the companies (Bezuidenhout and Buhlungu 2011). Spatial control regulates and restricts the movement of individuals or employees in industrial enclaves, worker accommodation facilities, industrial zones, or more broadly within a country (Kelly 2001). Reproductive or identity-based control intervenes in actual reproductive functions such as sexual intercourse, alimentation styles, or personal hygiene. Associational control or workplace containment regulate and curb worker organisations, trade unions, and sporting or religious activities (Bezuidenhout and Buhlungu 2011). Authoritarian control of production leads company management to create a state of fear among employees by overt displays of power (Kelly 2001). Gender-based control targets female employees in terms of labour allocation, career advancement and wage inequality (Kelly 2001).

Edwards (1982) offers two types of labour control, general directive controls and detailed control. While the former focuses on the extraction of surplus value, the latter is based on negotiation of order at work. Edwards argues that labour control results from the struggle between employers/managers and employees in which employers seek to minimise employees' opportunities for resistance while employees struggle for their rights. This can result in situations where employees acquiesce to domination by employers, a situation Edwards (1982) calls compliance. If, however, employees and employers manage to reach agreement, then they achieve consent.

Friedman (1977) offers two slightly different types of labour control: absolute control, in which the workers' productivity behaviours are directly controlled by hierarchical authorities, and relative control, where skilled workers are given responsibilities to control their subalterns.

Labour Reproduction

The last aspect of Peck's theory of social regulation is labour reproduction. With this concept Peck emphasises that all processes of labour incorporation, allocation and control relate to the household (family) and community. He argues:

> The sphere of production and the sphere of social reproduction are both separate and connected. They are separate in the sense that they each have their own structures of dominance along with their own distinctive rhythms and tendencies, but they are also related in the sense that each conditions and interacts with the other. (Peck 1996: 36)

Workers' productivity depends not only on their treatment in the workplace but also on the household, community and state that define the worker's social being. For Peck, research needs to look at how the reproduction of labour is regulated, including at how labour is trained and supported by family, community and state inputs outside the workplace. Likewise, it is crucial to know how labour is influenced by interconnected social realities such as the individual workplace, the industry or sector, and the locality, region or national economy, politics, culture and ideology.

Peck's theory, which is widely adopted by labour sociologists, is well suited for a careful analysis of the role played by all labour market actors in studies of post-war labour market reconstruction. We use Peck's theory for this purpose in this book because the concepts it proposes allow us to take into account the DRC's weak institutions and the large number of marginalised or unemployed people on the country's labour market. Equally important, Peck's theory allows us to understand how, historically and currently, the Congolese have been subjects of several contradictions, complexities and regulatory dilemmas. In short, Peck's theoretical framework is an important analytical tool for understanding how post-war labour markets are socially reconstructed.

Conclusion

This chapter elaborated concepts and theoretical perspectives that are applicable in the examination of the process of post-war labour market reconstruction. It first examined five key relevant concepts for examining the making, unmaking and remaking of labour markets in Southern Africa and the DRC in particular: the labour market, labour market institutions, the state, power and the international system. With regard to the term 'labour market', it explained that a comparatively narrow economic conception of the labour market has limitations for use in this book, while a broader, holistic view, advanced from various social science standpoints, has greater relevance for the arguments we make. As a post-war labour market occurs within a state, it is critical to understand the concept of the state from the perspective of the interpretations of the African state because these are sensitive to the complex situations that can be found in a former colony that has experienced multiple wars. Other approaches are too broad or too strongly based on labour situations of Western countries. Regarding Foucault's concept of power, it is important to use that concept for understanding activities of all actors involved in the Congolese labour market reconstruction. It is also useful to show how mechanisms of power and/or power relations shape processes of post-war labour market reconstruction. This also applies to the concept of the international system, which influences these processes in the current context of the post-Cold war era. With this, the chapter argued that Peck's theory of labour market social regulation provides an appropriate lens for examining processes of post-war labour reconstruction in Southern Africa and particularly the DRC. This choice was arrived at following the analysis of the merits and demerits of labour market segmentation theory. Our conclusion, therefore, is that Peck's theory was much more appropriate for the purposes of fulfilling the aims of this book.

References

Albritton, Robert. 1991. *A Japanese Approach to Stages of Capitalism Development*. New York: Palgrave Macmillan.

Aldrich, John. 1979. "Pigou's Wage-Good Method." *Recherches Économiques de Louvain/Louvain Economic Review* 45(4):317–50.

Alesina, Alberto, Michele Battisti and Joseph Zeira. 2018. "Technology and Labor Regulations: Theory and Evidence." *Journal of Economic Growth* 23(1):41–78.

Almeida, Rita K. and Jennifer P. Poole. 2017. "Trade and Labor Reallocation with Heterogeneous Enforcement of Labor Regulations." *Journal of Development Economics* 126:154–66.

Bachrach, Peter and Morton S. Baratz. 1962. "Two Faces of Power." *American Political Science Review* 56(4):947–52.

Barbieri, Paolo and Giorgio Cutuli. 2016. "Employment Protection Legislation, Labour Market Dualism, and Inequality in Europe." *European Sociological Review* 32(4):501–16.

Bassanini, Andrea and Ekkehard Ernst. 2002. "Labour Market Regulation, Industrial Relations and Technological Regimes: A Tale of Comparative Advantage." *Industrial and Corporate Change* 11(3):391–426.

Bayart, Jean-François, Stephen Ellis and Béatrice Hibou. 1999. *The Criminalization of the State in Africa*. Oxford: James Currey.

Berg, Janine. 2015. "Labour Market Institutions: The Building Blocks of Just Societies." Pp. 1–35 in *Labour Markets, Institutions and Inequality. Building Just Societies in the 21st Century*, edited by J. Berg. Cheltenahm, Northampton, Geneva Edward Elgar and International Labour Office.

Berg, Janine and David Kucera. 2008. "Labour Institutions in the Developing World: Historical and Theoretical Perspectives." Pp. 9–31 in *In Defence of Labour Market Institutions*, edited by J. Berg. London: Palgrave Macmillan.

Bezuidenhout, Andries, Christine Bischoff, Sakhela Buhlungu and Kezia Lewins. 2008. "Tracking Progress on the Implementation and Impact of the Employment Equity Act Since Its Inception." Vol. *Research Commissioned by the Department of Labour, South Africa*. South Africa: The Department of Labour.

Bezuidenhout, Andries and Sakhela Buhlungu. 2011. "From Compounded to Fragmented Labour: Mineworkers and the Demise of Compounds in South Africa." *Antipode* 43(2):237–63.

Boege, Volker, M. Anne Brown and Kevin P. Clements. 2009. "Hybrid Political Orders, Not Fragile States." *Peace Review* 21(1):13–21.

Bratton, Michael and Nicolas Van de Walle. 1994. "Neopatrimonial Regimes and Political Transitions in Africa." *World Politics* 46(4):453–89.

Braverman, Harry. 1974. *Labor and Monopoly Capital: The Degradation of Work in the Twentieth Century*. New York: Monthly Review Press.

Britwum, Akua O. and Leyla Dakhli. 2019. "Labour and the State." Pp. 495–522 in *General Labour History of Africa: Workers, Employers and Governments, 20th-21st Centuries*, edited by S. Bellucci and A. Eckert. New York: James Currey.

Brown, Deborah Wright and Alison M Konrad. 2001. "Job-Seeking in a Turbulent Economy: Social Networks and the Importance of Cross-Industry Ties to an Industry Change." *Human Relations* 54(8):1015–44.

Brožová, Dagmar. 2015. "Modern Labour Economics: The Neoclassical Paradigm with Institutional Content." *Procedia Economics and Finance* 30:50–56.

Buhlungu, Sakhela. 2010. "Trade Unions and the Politics of National Liberation in Africa: An Appraisal." Pp. 191–206 in *Trade Unions and Party Politics: Labour Movements in Africa*, edited by B. Beckman, S. Buhlungu and L. Sachikonye. Cape Town: HSRC Press.

Burawoy, Michael. 1979. *Manufacturing Consent: Changes in the Labor Process Under Monopoly Capitalism*. Chicago: University of Chicago Press.

Cai, Wenbiao and Manish Pandey. 2013. "Size-Dependent Labor Regulations and Structural Transformation in India." *Economics Letters* 119(3):272–75.

Cain, Glen G. 1976. "The Challenge of Segmented Labor Market Theories to Orthodox Theory: A Survey." *Journal of Economic Literature* 14(4):1215–57.

Chabal, Patrick and Jean-Pascal Daloz. 1999. *Africa Works: Disorder as Political Instrument (African Issues)*. Oxford: James Currey.

Clapham, Christopher S. 1985. *Private Patronage and Public Power: Political Clientelism in the Modern State*. London: Frances Pinter.

Clegg, Stewart R. and Mark Haugaard. 2009. "Introduction: Why Power Is the Central Concept of the Social Sciences." in *The Sage Handbook of Power*, edited by S. R. Clegg and M. Haugaard. Los Angeles: Sage.

Craig, Christine, Elizabeth Garnsey and Jill Rubery. 1985. "Labour Market Segmentation and Women's Employment: A Case-Study from the United Kingdom." *International Labour Review* 124:267.

Cramer, Christopher. 2008. "From Waging War to Peace Work: Labour and Labour Markets." Pp. 121–38 in *Whose Peace?: Critical Perspectives on the Political Economy of Peacebuilding*, edited by M. Pugh, N. Cooper and M. Turner. New York: Palgrave Macmillan.

Dahl, Robert. 1961. *"Who Governs?" Democracy and Power in an American City*. New Haven: Yale University Press.

Dahl, Robert A. 1971. *Polyarchy; Participation and Opposition*. New Haven: Yale University Press.

Dean, Mitchell. 2010. *Governmentality: Power and Rule in Modern Society*. New York and London: Sage.

Dimier, Véronique. 2010. *État et gouvernementabilité en Afrique*. Bruxelles: BESPO.

Doeringer, Peter and Michael J Piore. 1971. *Internal Labor Markets and Manpower Adjustment*. New York: DC Heath and Company.

Dugin, Alexander. 2018. "Multipolarity—The Definition and the Differentiation Between Its Meanings." *Multipolarity, Political Philosophy*. Retrieved 15th April 2018 (https://www.geopolitica.ru/en/1290-multipolarity-the-definition-and-the-differentiation-between-its-meanings.html).

Duverger, Maurice. 1988. *Sociologie de la politique: Éléments de science politique*. Paris: Dalloz.

Edwards, Richard. 1982. "Contested Terrain: The Transformation of the Workplace in the Twentieth Century." *Science and Society* 46(2).

Evans Jr, Robert. 1973. "The Rediscovery of the Balkans." *Keio Economic Studies* 10(1):33–38.

Fevre, Ralph. 1992. *The Sociology of Labour Markets*. London: Harvester Wheatsheaf.

Fine, Ben. 2003. *Labour Market Theory: A Constructive Reassessment*. New York and London: Routledge.

Foucault, Michel 1975. *Surveiller et punir*. Paris: Gallimard.

Freeman, Richard B. 2010. "Labor Regulations, Unions, and Social Protection in Developing Countries: Market Distortions or Efficient Institutions?" Pp. 4657–702 in *Handbook of Development Economics*, Vol. 5, edited by D. Rodrik and M. R. Rosenzweig. Amsterdam: Elsevier.

Freund, Bill. 1988. *The African Worker*. Cambridge: Cambridge University Press.

Freund, Bill. 2019. "Trade Unions." Pp. 523–52 in *General Labour History of Africa: Workers, Employers and Governments, 20th-21st Centuries*, edited by S. Bellucci and A. Eckert. New York: James Currey.

Friedman, Andy. 1977. "Responsible Autonomy Versus Direct Control over the Labour Process." *Capital & Class* 1(1):43–57.

Gartman, David. 1998. "Postmodernism; or, the Cultural Logic of Post-Fordism?" *The Sociological Quarterly* 39(1):119–37.

Giorgetti, Chiara. 2010. *A Principled Approach to State Failure: International Community Actions in Emergency Situations*, Vol. 64. Leiden and Boston: Brill.

Granovetter, Mark. 1985. "Economic Action and Social Structure: The Problem of Embeddedness." *American Journal of Sociology* 91(3):481–510.

Granovetter, Mark S. 1973. "The Strength of Weak Ties." *American Journal of Sociology* 78:1360–80.

Grimm, Sonja, Nicolas Lemay-Hébert and Olivier Nay. 2014. "'Fragile States': Introducing a Political Concept." Pp. 197–209, Vol. 35. Taylor & Francis.

Haltiwanger, John, Stefano Scarpetta and Helena Schweiger. 2014. "Cross Country Differences in Job Reallocation: The Role of Industry, Firm Size and Regulations." *Labour Economics* 26:11–25.

Hansen, Birthe. 2010. *Unipolarity and World Politics: A Theory and Its Implications*. London and New York: Routledge.

Harvey, David. 1989. *The Condition of Postmodernity an Enquiry into the Origins of Cultural Change*. Oxford: Basil Blackwell.

Harvey, David. 2007. *A Brief History of Neoliberalism*. Oxford: Oxford University Press.

Hasan, Rana and Karl Jandoc. 2013. "Labor Regulations and Firm Size Distribution in Indian Manufacturing." *Reforms and Economic Transformation in India*:15–48.

Hermet, Guy, Badie Bertrand and Philippe Braud. 1998. "Dictionnaire de la science politique et des institutions politiques." *Paris, Armand Colin (3e éd)*.

Hill, Christopher. 2015. *Foreign Policy in the Twenty-first Century*. Basingstoke: Palgrave Macmillan.

Howell, David R. 2004. *Fighting Unemployment: The Limits of Free Market Orthodoxy*. Oxford University Press.

Hudson, Kenneth. 2007. "The New Labor Market Segmentation: Labor Market Dualism in the New Economy." *Social Science Research* 36(1):286–312.

Hyde-Price, Adrian. 2007. *European Security in the Twenty-First Century: The Challenge of Multipolarity*. London and New York: Routledge.

Ikenberry, G. John, Michael Mastanduno and William C. Wohlforth. 2009. "Unipolarity, State Behavior, and Systemic Consequences." *World Politics* 61(1):1–27.

Inaka, Saint José 2020. "Post-War Labour Market Reconstruction: The Case of the Democratic Republic of the Congo." Doctoral dissertation, Sociology, University of Pretoria Pretoria.

Kalleberg, Arne L. and Aage B. Sorensen. 1979. "The Sociology of Labor Markets." *Annual Review of Sociology* 5(1):351–79.

Kaufman, Bruce E. 2007. "The Impossibility of a Perfectly Competitive Labour Market." *Cambridge journal of Economics* 31(5): 775–87.

Kelly, Mark G. E. 2010. *The Political Philosophy of Michel Foucault*. Routledge.

Kelly, Philip F. 2001. "The Political Economy of Local Labor Control in the Philippines." *Economic Geography* 77(1):1–22. https://doi.org/10.2307/3594084.

Keynes, John Maynard. [1935] 2018. *The General Theory of Employment, Interest, and Money. Introduction by Paul Krugman*. Cambridge: Springer.

Kreijen, Gerard. 2004. *State Failure, Sovereignty and Effectiveness: Legal Lessons from the Decolonization of Sub-Saharan Africa*, Vol. 50. Leiden and Boston: Martinus Nijhoff Publishers.

Krishnamurty, J. 2007. "Employment and Armed Conflict." *Indian Journal of Labour Economics* 50(1):47.

Langlois, Simon. 1977. "Les réseaux personnels et la diffusion des informations sur les emplois." *Recherches sociographiques* 18(2):213–45.

Lin, Nan, John C. Vaughn and Walter M. Ensel. 1981. "Social Resources and Occupational Status Attainment." *Social Forces* 59(4):1163–81.

Lukes, Steven. 2005. *Power: A Radical View*. New York: Palgrave Macmillan.

Marx, Paul. 2012. "Labour Market Dualisation in France: Assessing Different Explanatory Approaches." *European Societies* 14(5):704–26.

Mbembe, Achille. 2000. *De la postcolonie: essai sur l'imagination politique dans l'Afrique contemporaine*. Paris: Karthala.

Mpongo-Bokako, Bautolinga Edouard. 2001. "Institutions politiques et droit constitutionnel." Kinshasa: Editions universitaires africaines.

Paris, Roland. 2004. *At War's End: Building Peace After Civil Conflict*. Cambridge: Cambridge University Press.

Peck, Jamie. 1992. "Labor and Agglomeration: Control and Flexibility in Local Labor Markets." *Economic Geography* 68(4):325–47. https://doi.org/10.2307/144023.

Peck, Jamie. 1996. *Work-Place: The Social Regulation of Labor Markets*. New York: Guilford Press.

Pels, Dick. 2002. *Property and Power in Social Theory: A Study in Intellectual Rivalry*. London and New York: Routledge.

Phillips, Nicola. 2013. "Unfree Labour and Adverse Incorporation in the Global Economy: Comparative Perspectives on Brazil and India." *Economy and Society* 42(2):171–96.

Polanyi, Karl. 1944. *The Great Transformation*. New York: Farrar and Rinehart.

Reich, Michael, David M. Gordon and Richard C. Edwards. 1973. "A Theory of Labor Market Segmentation." *The American Economic Review* 63(2):359–65.

Rotberg, Robert I. 2004. *State Failure and State Weakness in a Time of Terror*. Washington: Brookings Institution Press.

Rotberg, Robert I. 2010. *When States Fail: Causes and Consequences*. Princeton and Oxford: Princeton University Press.

Rowe, Kelley. 2009. "Bottlenecks and Constraints within the Local Labour Market for Engineers in the Petrochemical Industry Sector: A Case Study of Engen Refinery, Wentworth." Master's dissertation, Development and Social Sciences, University of KwaZulu-Natal, KwaZulu-Natal.

Rubery, Jill. 2002. "The Labour Market Outlook and the Outlook for Labour Market Analysis." in *Changing Forms of Employment: Organizations, Skills and Gender*, edited by R. Crompton, D. Gallie and K. Purcell. London and New York: Taylor & Francis e-Library.

Rubery, Jill and Frank Wilkinson. 1981. *Outwork and Segmented Labour Markets*. London: Academic Press

Sadan, Elisheva. 2004. "Theories of Power." *Empowerment and Community, Planning*:32–71.

Saint-Paul, Gilles. 2000. *The Political Economy of Labour Market Institutions*. Oxford and New York: Oxford University Press.

Scott, David A. 2013. "Multipolarity, Multilateralism and Beyond...? EU–China Understandings of the International System." *International relations* 27(1):30–51.

Sender, John and Sheila Smith. 2003. *The Development of Capitalism in Africa*. New York: Taylor & Francis e-Library.

Spohr, Kristina and David Reynolds. 2016. *Transcending the Cold War: Summits, Statecraft, and the Dissolution of Bipolarity in Europe, 1970–1990*. Oxford: Oxford University Press.

Streeck, Wolfgang. 2005. "The Sociology of Labor Markets and Trade Unions." Pp. 254–83 in *The Handbook of Economic Sociology*, edited by N. J. S. A. R. Swedberg. Princeton: Princeton University Press.

Streeck, Wolfgang. 2010. "The Sociology of Labor Markets and Trade Unions." Pp. 254–83 in *The Handbook of Economic Sociology*, edited by N. J. S. A. R. Swedberg. Princeton: Princeton University Press.

Standing, Guy. 2011. *The Precariat: The New Dangerous Class*. New York: Bloomsbury Academic.

Thalakada, Nigel. 2012. *Unipolarity and the Evolution of America's Cold War Alliances*. New York: Springer.

Thomson, Alex. 2010. *An Introduction to African Politics*. New York: Routledge.

United Nations Treaty Collection, The UNTC. 2013. "The 1933 Montevideo Convention on Rights and Duties of States." *165* Montevideo: UN. Retrieved 28 May 2019, 2019 (https://treaties.un.org/pages/showDetails.aspx?objid=0800000280166aef).

Wachter, Michael L., R. A. Gordon, Michael J. Piore and Robert E. Hall. 1974. "Primary and Secondary Labor Markets: A Critique of the Dual Approach." *Brookings Papers on Economic Activity* (3):637–93.

Weber, Max. 1968. *Economy and Society; An Outline of Interpretive Sociology*. New York: Bedminster Press.

Williams, Eric. 2014. *Capitalism and Slavery*. UNC Press Books.

Wilthagen, Ton and Ralf Rogowski. 2002. "The Legal Regulation of Transitional Labour Markets." Pp. 233–73 in *The Dynamics of Full Employment: Social Integration Through Transitional Labour Markets*, edited by T. Wilthagen and R. Rogowski. Cheltenham and Northampton: Edward Elgar.

Wood, Geoffrey and Chris Brewster. 2007. *Industrial Relations in Africa*. Basingstoke: Palgrave Macmillan.

Yoder, Dale and Herbert Heneman. 1959. *Labour Economics and Industrial Relations*. Chicago: South-West Publication.

You, Jong-Il and Ha-Joon Chang. 1993. "The Myth of Free Labour Market in Korea." *Contributions to Political Economy* 12(1):29–46.

Zoellick, Robert B. 2008. "Fragile States: Securing Development." *Survival* 50(6):67–84.

PART II

Labour Markets and Post-War Labour Market Reconstruction in the SADC Region: Insights from the Democratic Republic of the Congo

CHAPTER 4

Dynamics in the Making of Labour Markets

We start engaging with the empirical aspects of the making, unmaking and remaking of labour markets in Southern Africa in this chapter and narrow down to examine these realities in a SADC member state. We argue that, state reforms were not sufficient to facilitate post-war labour market reconstruction in the SADC member state in question. The argument is based on textual evidence and data from personal face-to-face interviews and focus group discussions with selected respondents in the Democratic Republic of the Congo (DRC). The argument is also couched in the conceptual and theoretical perspectives outlined in Chapter 3, and as applied to the historical development of the labour market in the DRC. In the current chapter, we also trace the past of the DRC's labour market and its contribution in shaping the market in the twenty-first century. We analyse the implementation of the forced labour regime in the Congo Free State (1885–1908) and the racialised dual labour market in the Belgian Congo (1908–1960). We then show how wars in the country balkanised its post-independence labour market (1960–1965) and, finally, show how the market went through an endless process of regress during President Mobutu's presidency (1965–1997) and Laurent-Désiré Kabila's (1997–2001) presidency following the 1996–1997 war. Further, we examine the factors that conditioned and inhibited attempts by the state to achieve post-war labour market reconstruction between 2003 and

S. J. Inaka et al., *The Reconstruction of Post-War Labour Markets in The Southern African Development Community*,
https://doi.org/10.1007/978-3-031-37357-2_4

2019 under President Joseph Kabila (2001–2019). Here we show that the power-sharing deal brokered during Joseph Kabila's presidency negatively impacted the reconstruction of the post-war labour market in the DRC. This is because it hindered the reconstruction of the market during the country's transition (from 2003 to 2006) into a purported democratic dispensation (Tshimpaka and Nshimbi 2022). Despite the adoption of a new constitution in 2006 and election of new governments in 2006 and 2011, the implementation of labour reforms in the DRC was significantly delayed and this was at the expense of the labour force. We also use statistical data to augment our analysis of the DRC's failure and the challenges it faced in reconstructing the labour market.

Accordingly, this chapter is divided into six sections, which consider the DRC's forced labour regime under the Belgian King Leopold II (1885–1908); the dual labour market regime (1908–1960); fragmented labour (1960–1965); the Mobutist labour market (1965–1997); the post-Mobutist labour market (1996–2002); the challenging post-war period of labour marker reconstruction (2003–2018); and lastly, statistical data that helps to describe the challenges of labour market reconstruction in the DRC's post-war labour market.

King Leopold II's Forced Labour Regime (1885–1908)

The colonisation of the Congo started as a personal and private business venture of Belgian King Leopold II from 1885 to 1908. He established a ferocious forced labour regime through three key strategies: land dispossession, labour taxation, and the use of violence. He deployed a set of draconian laws to dispossess the people of their land (Au 2017). The Land Occupation Ordinance, the Royal Decree on Land Ownership, and the Land Demarcation between the state and other owners facilitated the land dispossession process (Inaka 2020). Many people lost livelihoods as a result and were exposed to Leopold II's forced labour regime (Nzongola-Ntalaja 2002; Ewans 2003).

The Congo Free State under Leopold II was ruthless and violent, especially during worker recruitment campaigns, and aimed at ensuring control (Hochschild 1998). The regime captured and conscripted Congolese into gathering rubber and ivory. It deployed disproportionate military force, involving burning down villages, looting, raping women, killing and mutilation, to enforce this unfree recruitment of labour (Ahmad and Awan 2017). It was further characterised by unpaid labour on call 365 days a year (Daniels 1908), corporal punishment (Ewans 2003), and used death by torture to instil fear. Male workers were, for instance, tied to trees and their genitals, right hands and heads mutilated if they did not bring in required quotas of rubber (Reybrouck 2014).

Leopold II's labour regime triggered the first global human rights campaign in the twentieth century, which explicitly drew on the language of crimes against humanity (Weisbord 2003). The campaign was led by English-speaking activists like Joseph Conrad, George Washington Williams, Roger Casement, Edmund Morel, and the Congolese Reform Association. It pressured the British and American governments to intervene. In 1908, Leopold II gave in to the pressure and ceded ownership of the Congo Free State to the Belgian government. From 1908 to 1960, the Congo Free State became Belgian Congo (Hochschild 1998).

This brief history demonstrates that the thirst for African raw materials encouraged the establishment of a labour market driven by forced labour. The multipolar international system that characterised the early stage of colonialism influenced the installation of forced labour regimes in many Southern African countries, including the Congo. But these regimes were subsequently resisted by Africans and international activists, who deployed multidirectional power at micro and macro levels (Foucault 1975).

Labour Market Regime Under Belgian Colonial Rule (1908–1960)

A dual labour market regime (*régime de dualité*) replaced the forced labour regime of the Congo Free State when the Belgian government replaced King Leopold II's reign. The new regime was equally unequal, racialised and gendered. It had differentiated privileges for white colonialists from underprivileged Congolese. Belgian workers worked in a free (even democratic) environment while Congolese were subjected to forced labour. The dual regime guaranteed a ready supply of African labour in quantities that capitalist investors required. The colonial administration further developed four labour market policies—rationalisation of the workforce, compulsory cultivation, paternalism and '*Belgianisation*'—to buttress the system and guarantee a cheap, available supply of Labour for the administration and private companies.

Labour incoporation strategies within the rationalisation policy sought to end the forced labour regime and introduce a waged labour market that would (still) force Congolese to supply labour to the waged labour market. On 1 July 1910, for example, the Belgian government passed a Poll Tax Decree that introduced a waged labour market in the Congo (Vaessen 2001). This compelled Congolese to work for cash (in order to pay tax) in economic sectors dominated by Europeans (Houben and Seibert 2013).

In 1917, the colonial authorities partially restored forced labour through the policy of compulsory cultivation. In 1933, it established workloads for educational purposes through the '*Les travaux d'ordre educatif*' (TOE) policy. However, rationalisation and compulsory cultivation policies induced Congolese to seek jobs in areas where white employers needed them. Both systems remained in force until 1960 (Soriano 2017). King Leopold II's legacy in the labour market continued too, notwithstanding the new policies.

The policy of paternalism kept the labour force under control and ensured labour was reproduced, despite being implemented to promote the well-being of workers (Reybrouck 2014). The rationale was to encourage and develop employees' sense of belonging to companies, at a time when rapid industrialisation in the country made permanent cheap labour necessary (Bezy 1957). The colonial state and private companies intended to have this strategy to keep workers under control. It restricted the workers' geographical mobility and only allowed them to move with permission from state authorities and employers. The policy also contributed to the reproduction of labour, as employers assumed responsibility over employees' welfare and that of their families including the provision of accommodation, health and leisure and education. The policy also banned married women from participating in the labour market without the written permission of their husbands (Gondola 1997).

In contrast to these policies that targeted Congolese, the colonial state enacted a *Belgianisation* policy that favoured and gave privileges to Belgians in the labour market. The policy included a set of economic and labour allocation measures that increased the percentage of Belgians in the total white population, the monopoly of Belgian investment and access to the colony's labour. The policy was designed to address fears that Belgium might be side lined and outcompeted in the ownership of the Congo by powerful countries (Vaessen 2001; Abbeloos 2008).

The Belgians could not implement the *Belgianisation* policy at first because of local and international criticism. Britain pressured Belgium to instead enact a labour contract act similar to the acts in British colonies. The Belgian authorities, therefore, passed a labour contract act on 17

August 1910. But companies and some governors in regions of the colony that depended on agriculture considered the act unworkable. They argued that the legislation would prevent them from treating their workers according to what the new legislation stipulated. Consequently, the act was only enforced in the mining sector. The Belgian colonial regime selectively enforced the labour contract act until the end of WWI (Vaessen 2001). The *Belgianisation* policy on the other hand became effective after WWI, especially in the 1930s when the Belgian government reasserted direct control of the economy and workers (Manwana 1987).

The colonisers also passed a set of racialised labour laws in the 1920s and 1930s. The Labour Contract Act between 'Civilised Masters' and 'Indigenous Workers' of 16 March 1922, for example, introduced new regulations for the relationship between black employees and white private employers (Congo Belge 1922c) while the Royal Decree of 13 March 1922 addressed the relationship between white employers and white employees in the private sector (Congo Belge 1922d). The public sector was regulated by two decrees. While the Royal Decree of 5 February 1922 on the status of officials and agents of the colony (Congo Belge 1922a) was directed at European public servants, the Royal Decree of 27 February 1922 on the status of African personnel of the colony was aimed at Congolese public servants (Congo Belge 1922b). As the titles suggest, the labour laws established a colour-bar and gendered discriminations in the labour market where they gave white employees more privileges than Congolese employees.

The labour legislation enacted by the Belgian colonial regime created a dual labour market that privileged white over black employees. Labour recruitment processes favoured European public servants from Belgium and Luxemburg who were trained in Belgium. These workers were offered higher employment grades in the Congo. Their lowest positions were the higher grades for Congolese officials (Brausch 1961). Differences between underpaid Congolese and well-paid European employees matched these labour inequalities (Manwana 1987). Unfree colonial labour relations characterised by the prohibition of trade unions imposed limitations on Congolese ability to claim their labour rights (De Clerck 2006).

Institutions and actors like the *Office des Affaires Indigènes et Main d'œuvre* (Indigenous Affairs and Labour Office), private companies, faith-based organization (FBO)-run and public schools, traditional chiefs, formal and informal labour brokers and workers' organisations regulated the dual labour market. The Indigenous Affairs and Labour Office managed supply and demand in the labour market from 1910 to 1960 (Lwamba 1985). Private companies played a key role on the demand side as primary employers of Congolese workers (Bezy 1957). On the supply side, FBO-run and public schools controlled the educational system and thus supplied workers to the state and private companies. Other labour suppliers were traditional leaders and labour brokers. The colonial government appointed 'traditional chiefs', who were seldom from noble lineages but rather docile individuals who were loyal to the government, to supply able-bodied men for the workforce. Similarly, formal or informal labour brokers, mostly private contractors and white adventurers, used violent labour recruitment methods (Lwamba 1985; Young and Turner 1985; Reybrouck 2014).

Other players in the labour market included professional organisations which were officially recognised as trade unions in 1946. Trade unions were active in the labour market and the political struggle against colonisation in the late 1950s (Lwamba 1985). As more trade unions received international support from pro-communist trade unions in Europe, the colonial administration placed trade unions in the country under the tutelage of Belgian union representatives as a measure of mitigating the spread of communism in the country through unionisation. Nonetheless, these unions and political leaders fought against colonisation and the racialised labour market (Young and Turner 1985). Although some features of the colonial labour market like discrimination, anti-unionisation practices and the unfair treatment of employees persist, the dynamics of the bipolar international system after WWII and local resistance to colonial rule brought the dual labour market and colonisation in the Congo to an end in 1960.

Post-independence Labour Market Fragmentation (1960–1964)

The labour market experienced fragmentation between 1960 and 1964. This was because of three attempted secessions between 1960 and 1963 and two rebellions in 1964. A key driver of this was the dual labour market inherited from the Congo's colonial past. Just five days after independence, on 5 July 1960, Congolese soldiers mutinied in Leopoldville (now Kinshasa) and other major cities. A Belgian Commander-in-Chief of the *Forces Publiques*,[1] General Janssen's refusal to give leadership roles in the army to nationals triggered the mutiny (Leslie 1993). The soldiers engaged in violent demonstrations when they realised that their positions would not change (Haskin 2005).

The first secession attempt occurred eleven days after Independence Day, on 11 July 1960. This was when Governor Moise Tshombe declared Katanga Province an independent state. The province had an administrative system similar to that of a state, its own currency, a flag, national anthem and an organised labour market (Larmer and Kennes 2014). Its labour market was based on regionalist principles that promoted the participation of Katangese people only (Kairouz 2016). The labour market ended in January 1963, when a United Nations (UN) military intervention thwarted the Katanga secession and saw Moise Tshombe go into exile. The peace negotiations that ensued ushered Tshombe into the premiership on 6 July 1964. However, President Kasa-Vubu dismissed Tshombe on 13 October 1965 and he subsequently went into exile in Spain. Tshombe was eventually arrested, tried and imprisoned in Algeria where he remained until his death in 1969 (Kisangani and Bobb 2010).

The second secession attempt involved South Kasai, which seceded in 1960 under Albert Kalonji's leadership. Kalonji introduced a mono-ethnically based labour market regime, where only Luba speaking people of the region had the right to work (Nzongola-Ntalaja 2002). Though traditional chiefs supplied Luba speaking people only to the workforce, the militaristic and unfair labour treatment of these employees affected the region's tribalized labour market until Kalonji's downfall in 1962 (Parkham 1992). Kalonji was ousted from Kasai in 1962 because of conflicts between secessionist leaders and was compelled to join the government in 1964 (Young 1965; Kisangani and Bobb 2010).

The third attempt was when the former Oriental Province seceded to form the Free Republic of the Congo in 1961. Antoine Gizenga led that secession attempt. He centralised authority and interfered in the hiring and firing of public officials. This created tension between his government and senior executives of the public administration until the end of his secession attempt in 1962 (Young 2015). Although Gizenga joined the government in 1961 after negotiations for peace, he was charged with

[1] This was a gendarmerie and military force throughout the period of Belgian colonial rule.

treason, imprisoned from 1962 to 1964 but reprieved and went into exile from 1965 to 1990 (Kisangani and Bobb 2010).

The pro-Lumumba *Front pour la Libération Nationale du Congo* (Front for the National Liberation of Congo, FLNC) attempted a quasi-communist rebellion against the state in 1964. Christophe Gbenya, Gaston Soumialot, and Laurent Désiré Kabila led the rebellion. The rebels invaded almost half of the DRC covering the vast eastern part of the country. The rebellion disrupted the labour market. Their ally Pierre Mulele controlled a major area in the Southern-Western part of the country in Kwilu Province. Despite the support of the Soviet Union, China and Cuba, the *Simbas* or lions, as the rebels were known, could not organise a structured administration or labour market but total anarchy and ferocity against public workers and civilians (Young 1965). However, a collusion between the army, mercenaries and Belgian soldiers defeated these rebels and their leaders went into exile (Young and Turner 1985; Haskin 2005; Kisangani and Bobb 2010).

The three secessions and the rebellions disrupted the labour market since most of the institutions could not properly function. Effectively, the situation rendered itself to four 'governments' and four labour markets. The country witnessed high levels of militarisation, politicisation, ethnic division, regionalism, discrimination and weak labour market institutions. Even if President Joseph Kasa-Vubu in Leopoldville (Kinshasa) was the official leader of central government, the country fell under the control of General Joseph Mobutu.

Two notable continuities and discontinuities in the labour market of the 1960s are worth highlighting. First, the racial labour discrimination of the colonial period transformed into ethnic, regionalist, neo patrimonial labour discrimination. This has persisted to the present day as the rest of this book shows. People in different contexts have reproduced discrimination based on the ascribed statuses of individuals—i.e. political, regional, or ethnic identities. Second, the convergence of a combination of factors including the bipolar international system that pitted communism against capitalism, Belgian neo-colonialism, internecine conflicts among local elites in the country, quasi-improvised independence processes, rich natural resources and the strategic interests of foreign powers in the country fuelled the political crisis in the country. More relevant to the discussion in this book was the lack of a trained cadre of Congolese in the post-colonial public administration until the international community, led by the UN, deployed skilled technicians, professionals and specialists

(Ndaywel 1998). These efforts helped improve labour supply and were particularly visible in the Mobutu era.

Labour Market Under President Mobutu (1965–1997)

Mobutu ruled over a largely authoritarian, mismanaged, corrupt and kleptocratic regime from 1965 to 1997 (Bach 2011). After a brief amelioration from 1967 to 1973, the country's labour market deteriorated, especially from 1973 to 1997. Various factors, actors and processes contributed to this.

The Golden Age (1965–1973)

General Joseph Désiré-Mobutu (1930–1997), Commandant-in-Chief of the army, ousted President Joseph Kasa-Vubu in a coup d'état on 24 November 1965. Mobutu implemented a series of political and economic reforms between 1965 and 1974 that impacted the labour market. His regime enacted the Bakajika Act on 7 June 1966. This land act stipulated that all the wealth above and below the ground belonged to the state. This 'formally extinguished all land grants and concessionary powers delegated by the colonial state' (Young and Turner 1985). The regime also formed the *Mouvement Populaire de la Révolution* (the Popular Movement of the Revolution, MPR), the ruling party, on 20 May 1967. The MPR and 'administrative responsibilities were literally fused together, thereby extending the role of the party to all administrative organs' including the *Union Nationale des Travailleurs du Congo* (National Union of the Workers of the Congo) (Haskin 2005). Mobutu grouped all labour market institutions (colonial corporations, trade unions, schools, universities) under his centralised, patrimonial, and authoritarian regime (Young and Turner 1985).

These political reforms set the ground for some major economic reforms aimed at nationalising the economy for economic growth. First, the nationalisation of the *Union Minière du Haut Katanga* (Mining Union of the Upper-Katanga, UMHK), a major mining company and employer of Congolese workers, on 1 January 1967. The UMHK was renamed the *Generale Congolaise des Minerais* (Gocomin), and, later, *Générale des Carrières et des Mines* in 1971 (Leslie 1993). Second, the monetary reform on 24 June 1967. This freed the Congolese currency from the Belgian franc. The creation of a new currency, called Zaïre, resulted in monetary stability, resumption of economic growth and wage increases (Bossekota and Sabiti 2001). Third, the enactment of the Congolese Investment Code on 10 February 1969. This attracted investors (Young and Turner 1985) and increased job opportunities in the early 1970s (Ndaywel 1998).

Together, the cited economic and political reforms in turn laid the ground for the reform of the labour market during Mobutu's reign, leading to several labour market policy reforms. The first of these reforms of the labour market was oriented to the control of workers and trade unions. The creation of the *Union National des Travailleurs Congolais* (Congolese National Trade Union, UNTC), a single trade union to represent both public and private sectors, in June 1967 undermined the unions' main role of protecting members' interests. Effectively, the UNTC engaged in collective bargaining by becoming part of the state apparatus (Young and Turner 1985).

The second labour market policy reform included the implementation of the *Salongo* in 1968. This continued until 1990. *Salongo* consisted in compulsory 'civil service', wherein every citizen had to work an afternoon a week. This usually happened on Saturdays and included agricultural, public cleaning and 'development' work (Soriano 2017). According to Haskin (2005: 66) '*Salongo* was ironically reminiscent of the earlier colonial forced labour'.

The third reform focused on the private sector. The Congolese Labour Code enacted on 9 August 1967 facilitated the protection of employees in this sector. The state subsequently came up with the *Service National de l'Emploi* (National Employment Service, SENEM) on 29 October 1967 to guide the task of recruiting workers previously performed by formal and informal labour brokers. However, SENEM was not a successful institution (Ilofo 1973). Politicians, academics, army officers, religious leaders and Mobutu's relatives and cronies continued to unofficially and 'privately' recruit and hire workers (Inaka 2020).

The fourth reform concerned the functioning of the public administration and recruitment and hiring procedures of public servants. The Ordinance Law No. 72/023 of 4 July 1972 on the status of public servants made the public sector subservient to the one-party state. It stipulated that people seeking employment in the public sector had to pass entrance examinations. This 'reform' placed all procedures for the recruitment and hiring of public servants in the hands of a small commission. This politicised the public administration, as a few influential members of the one-party state became the main employers (Mpinga and Gould 1975).

The last reform involved the creation of the *Association Nationale des Entreprises du Zaïre* (National Association of Zairian Enterprises, ANEZA) on 14 September 1972, an employers' organisation and chamber of commerce set to protect members' interests and promote private sector activities in the Congo (Matenda 2002). ANEZA was created to replace the colonial employers' organisations that previously controlled the economy with a new one through which Mobutu could keep corporates in check (Williams 2014).

The Mobutu government thus centralised the labour market under the management and surveillance of one political party. This restored order from the post-independence fragmented labour market (1960–1964). However, some labour market practices like anti-trade unionisation or coteries in hiring processes persisted into the early Mobutu era. A closer look actually shows that the Mobutu regime drew quite widely on the colonial period in defining its politics and labour market regime. Not only so, the improved functioning of the labour market was short-lived: the labour market was about to disintegrate.

Disintegration (1974–1997)

A series of events led to the disintegration of the labour market in the DRC from 1974 to 1997. The misconceived and poorly implemented '*Zaïrianisation*' Mobutu announced on 30 November 1973 contributed to the disintegration. The policy sought to transfer the DRC's wealth from foreign ownership to Congolese. This led to the expropriation of almost 2,000 foreign-owned companies and the transfer into the ownership of local elites (Haskin 2005). The coterie of about 300 individuals around the president who seized most of these companies were mostly incompetent in business management and within six months most of the companies had gone bankrupt. *Zaïrianisation* directly affected the labour market because it led to salary arrears, deductions and unfair summary dismissals (Young and Turner 1985; Bossekota and Sabiti 2001). Unexpected international and national events like the 1973 oil crisis and the collapse of international copper and cobalt prices destabilised the economy and labour market too (Young and Turner 1985).

These labour market realities precipitated two interrelated phenomena in the 1970s. First, job seekers and employees who sought promotion turned to non-market actors, called *des parrains* (godfathers or sponsors), for assistance (Lumbu 2004). *Des parrains* included high-ranking army officers, politicians, professors, Mobutu's siblings and other influential persons who shared ethnic, provincial or religious affiliations (Ndaywel 1998). Second, public sector workers staged wildcat strikes (Mwabila 1979), which Mobutu's army often brutally repressed (Leslie 1993). The labour market thus deteriorated and continued to do so into the 1980s and 1990s.

The Mobutu regime failed to mitigate the economic challenges in the DRC. In 1983, the IMF forced him to implement Structural Adjustment Programmes (SAPs). But the SAPs did not improve living standards or the labour market. A significant percentage of the state's revenue went into paying off external debt. Consequently, the privatisation of public companies that came as a condition for IMF support only shrunk the job market (Kawu 2013). The labour market saw downsizing and massive dismissals (Bossekota and Sabiti 2001), unpaid salaries and an increase in informal employment (De Boeck 2004).

The Mobutu regime also enacted Law No. 81/003 of 17 July 1981 on the Status of Civil Servants. This law stipulated that people seeking work in the public sector had to pass a written examination (Tshombe 2017). But this requirement was not strictly applied, with state authorities and non-market actors continuing to hire based on patronage, ethnicity, or corruption (Kawu 2013).

Political and economic conditions in the DRC worsened in the 1990s. The new post-Cold War world and economic order and deteriorating political relations between the USA and Zaïre weakened the Mobutu regime. Locally, opposition parties pressed for democratic reforms in the 1990s (Nzongola-Ntalaja 2002). The economy also collapsed in the 1990s, with overlapping effects on the labour market. Corruption, kleptocracy and economic maladministration and an unsafe business climate leading to a decrease of production in the mining industry after decades of underinvestment (Kawu 2013). Besides endless inflation (Bossekota and Sabiti 2001), looting in many cities in 1991 and 1993 wrecked the few remaining companies and led to further unexpected job losses (Trapido 2016).

The dysfunctional labour market actors and institutions in the Zaïre of the 1990s reflected the image of a politically, socially and economically failing country. Mobutism was a source of Zaïrian pride in the early 1970s that became a sarcastic term for malpractice in the country (Kisangani and

Bobb 2010). Though the Mobutu regime set up several labour market institutions, most became empty shells that finally collapsed under the stewardship of Laurent-Désiré Kabila.

Labour Market After Mobutu's Reign (1997–2002)

Laurent-Désiré Kabila's authoritarian regime tried to reconstruct the labour market after Mobutu's regime collapsed. The labour market, however, remained poorly regulated for various economic reasons, the persistence labour market features established during Mobutu's reign, and the war of 1998–2002.

Kabila's regime failed to build a well-structured democratic state. He ruled the DRC by decrees, which enabled him to co-opt members of parliament, arrest opponents and restrict opposition political parties (Reyntjens 2009). He used ethnic and patron-client networks to appoint Congolese and Rwandan Tutsis, Congolese businessmen, intellectuals in the diasporas, former politicians from the Mobutu regime, people from Katanga and cronies to key positions in state institutions (Stearns 2012). The appointees, in turn, hired their kinsmen in lucrative job positions (Reyntjens 2009). At the same time, the Kabila regime manipulated, frightened, or silenced trade union leaders. The weakening of trade unions encouraged public companies to use unfree labour, as public authorities could hire and fire at will (Oppong 2007). For instance, on 6 November 1998 Kabila himself arbitrarily dismissed 315 magistrates (Basila 2004) and appointed judges at will (Oppong 2007). His regime failed to reconstruct the post-war labour market of the country because of the ruin caused under Mobutu and because of the 1998–2002 civil war. Nine African countries and five rebel movements got directly involved in the latter war and this resulted in six different labour markets in the country. The involvement of multiple actors in that war and threats to international security made it regionally imperative for the war to end. The international community compelled the warring factions to sign a series of peace agreements and imposed peace, democratisation and economic liberalisation in line with neoliberal norms (Marriage 2011).

The process started with the signing of a ceasefire agreement in Lusaka on 10 July 1999. Several peace-making moves followed including, inter alia, inter-Congolese dialogue, the establishment of a disarmament mechanism, and the creation of a UN peacekeeping force, the *Mission des Nations Unies en République démocratique du Congo* (United

Nations Mission in the Democratic Republic of the Congo, MONUC) (Nzongola-Ntalaja 2004). The Lusaka agreement, however, yielded limited results because the warring factions failed to uphold it up to Kabila's assassination on 16 January 2001. His son, Joseph, replaced him as president and facilitated an inter-Congolese dialogue process in Sun City, South Africa, from 25 February to 19 April 2002 (Trefon 2011).[2] The dialogue led to the signing of the Global and *All-Inclusive Agreement on the Transition in the Democratic Republic of the Congo* or the Pretoria Agreement in Pretoria on 17 December 2002. This offered new hope for transition and democracy in the DRC and provided the foundation for a new national constitution. The transition would be based on the sharing of military, political and socioeconomic power between the warring factions, civil society and unarmed opposition (Inter-Congolese Dialogue 2002). The power-sharing deal lasted from 2003 to 2006. The core goals of the transitional government were to restore government authority; foster national reconciliation; improve security; organise elections; and create new political institutions (Trefon 2011). The transition period (2003–2006) marked the beginning of a challenging era of post-war labour market reconstruction in the DRC.

Post-war Labour Market Reconstruction After the Year 2000

The state attempted to reconstruct the labour market after the devastation of the 1998–2002 war in the DRC. Its efforts are evident in two time periods including, the transition, stretching from 2003 to 2006, and the end of that period up to 2018. Despite unifying the DRC through a power-sharing arrangement, the 2002 Pretoria Agreement had unforeseen impacts on post-war labour market reconstruction. This is because the agreement was an outcome of a political process managed from outside the DRC's sociocultural, political and economic realities (Trefon 2010). It produced a patrimonial state dominated by political elites associated with President Mobutu or President Laurent-Désiré Kabila (Autesserre 2010). Labour market conditions in the DRC during

[2] On the 16 January 2001, the President Laurent D. Kabila was killed by his bodyguard. Then, the Minister of information officially announced that the Congolese government found it convenient to designate Joseph Kabila as the successor of his father (see Nzongola-Ntalaja 2004).

the period of transition can be understood through three broad themes including power sharing and the labour market, labour reforms in the private sector and labour reforms in the public sector.

The power-sharing regime of 2003–2006 politicised the public sector labour market by establishing multiple chains of command in the various political parties involved. This not only rewarded former enemies with access to power and jobs but also adversely affected the functioning of the market (Autesserre 2010). Regarding the latter, Kodi (2008) says about the public administration and public companies that,

> all the signatories were allocated a proportionate share of government ministries, seats in the national assembly and the Senate, top positions in the state enterprises, ambassadorial positions, and high-ranking positions in the army and police. The appointments to these positions were carried out by the signatories to the Agreement, i.e. the President and the four Vice-Presidents. (Kodi 2008)

The people appointed in this manner were only accountable to their recruiters/appointers and not the country's public administration. Political patronage took precedence over merit in official appointments (Kodi 2008). The power-sharing arrangement had a further four major and interlinked implications for labour market regulation during the transition period. It politicised the public labour market; contributed to contradictions and overlaps between the transitional constitution and existing labour laws; generated multiple chains of command in the public labour market; and increased political patronage, nepotism, ethnic and gender discrimination and corruption in processes of labour incorporation, allocation and control in the public labour market.

Regarding the politicisation of the labour market, former enemies exploited the principles of the Pretoria Agreement to wield power and authority in the public labour market. They had this conviction that they had the right under the agreement and transitional constitution to recruit and fire public employees. Politicians and high level bureaucrats thus exploited the Pretoria Agreement and the transitional constitution to appoint and dismiss public workers at will.

The contradictions and overlaps in the transitional constitution and Act No. 81/003 of 17 July 1981 on the Status of Civil Servants contributed to confusion in the hiring and management of public officials. While the transitional constitution allocated the highest positions to the former enemies and confirmed their right to recruit, appoint and manage employees, Act No. 81/003 stipulated that only candidates who had passed a written examination may be recruited to public administration positions and that promotions should be based on merit, seniority, performance, integrity and outstanding achievement (République du Zaïre 1981; République Démocratique du Congo 2003). The parallel applicability of these two legislations caused tensions between the former adversaries over job recruitments and appointments (Ngoy-Kangoy 2006).

The multiple chains of command in the public administration negatively affected the implementation of the Mbudi Agreement on the minimal wage. The transitional government had entered into the Mbudi Agreement with the trade unions on 12 February 2004 and pledged to pay the lowest paid public service employee US$208 and the highest paid employee US$2,080 (République démocratique du Congo 2004). However, then national Vice-President, Jean-Pierre Bemba, declared that it was impossible for the Economic and Finance Commission to respect the Mbudi Agreement because of budgetary constraints. The wage policy was, therefore, never implemented. During a personal face-to-face interview, a Congolese Sociologist made the following observation on the consequences of non-implementation of the Mbudi Agreement.

> Remember the Mbudi Agreement with Vice-President Z'ahidi Ngoma, they deliberately refused to implement them. They found unconvincing excuses, like the country can't afford to pay $250 for the lowest employees. ... Look, today an MP gets $13,000 a month, but a policeman gets $50. So the policeman has to work for more than 20 years in order to reach what an MP receives each month. (Sociologist, Personal Interview, University of Kinshasa, April 2017)

The Mbudi Agreement shows the extent to which the transitional constitution created conflicts of jurisdiction and contradictions within the transitional government. Ngoma used his constitutional prerogative to sign the Mbudi Agreement while Bemba used his own constitutional prerogative to reject this decision.

The power-sharing agreement also helped create conditions that made clientelism, nepotism, tribalism, and corruption[3] central management

[3] Corruption in the public labour market was enacted in two ways in particular. In what was colloquially called an *opération retour* ('return operation'), a person would be

strategies in the public labour market. In the private labour market they led to mass dismissals and downsizing and outsourcing of labour.

In a face-to-face interview, a former minister bemoaned the effects of a politicised labour market and expressed concern over the absence of meritocracy in hiring and appointing employees:

> Many uneducated persons, villagers — without military training — became generals, colonels, majors in our army … . So, it was just *la tour de Babel* [a bewilderingly confusing situation] in the army … Politicians [and] military careerists obliged state officials to hire [and] promote members of their families, their clients, their comrades, and so on. (Former minister, Personal Interview, Kinshasa, July 2017)

A human rights activist expressed particular concern with the opportunity for sexual harassment that arose from this and especially the practice of *promotion canapée*[4]:

> That *promotion canapée* was actually at its highest level. Politicians used our young ladies. They slept with them and gave them useless job positions. Sir, let me tell you that politicians messed up with women during the transition. (Female human rights activist, Personal Interview, Kinshasa, February 2017)

Employees experienced different realities in the private labour market. Our data suggest significant mass dismissals in private companies in the period 2003–2005. An economist explained the mass dismissal of employees in the private sector in relation to the capitalist nature of wealth accumulation and the weak government institutions in the country:

> Companies seek maximum profits. It wasn't their fault. … As our institutions were weak, they took that opportunity. That's how capitalists act. (Economist, Personal Interview, Free University of Kinshasa, April 2017)

hired if he/she agreed to pay 10% of the salary to the person who hired him/her. And in a *promotion canapée* (with a canapé being a decorative sofa), a woman will be hired or promoted if she agrees to grant sexual favours to the person hiring her.

[4] In a promotion canapée (with a canapé being a decorative sofa), a woman will be hired or promoted if she agrees to grant sexual favours to the person hiring her.

On radical downsizing, a unionist was certain that there were reciprocal exchanges of favour between politicians and some companies implementing these labour measures:

> We tried convincing the government to take a strong decision against that downsizing. … Nothing happened as many authorities were in cahoots with employers. (Unionist, Personal Interview, Kinshasa, December 2016)

The DRC constitution implicitly emptied labour market institutions of their customary authority on matters of recruitment, hiring and management of personnel.

The labour market in the private sector was not spared from the impacts of the power-sharing regime either. Politicisation of the labour market in that sector contributed to mass dismissals, outsourcing of labour and downsizing in the private sector labour market. Just before the transitional period, parliament amended two laws—the 1967 Labour Code and the 1969 Investment Code. The power-sharing regime endorsed both laws (Moshonas 2014). International actors were crucial to both pieces of legislation and they both significantly impacted the implementation of reforms in the private labour market.

Government enacted Investment Code Act No. 2002/04 of 21 February 2002 to attract private investment (République démocratique du Congo 2002b). The Code led to the passing of Decree No. 2002/065 of 5 June 2002 which established the National Agency for the Promotion of Investment (ANAPI) (République démocratique du Congo 2002a). Parliament also enacted the Labour Code Act No. 2002/015 of 16 October 2002, to regulate the private labour market (République Démocratique du Congo 2002c). The act created ONEM as the official intermediary between private employers and jobseekers, to provide legal protection for the national workforce vis-à-vis foreign competition (van Impe 2012), and expand labour legislation to cater for informal and formal employees in all sectors—including those in sociocultural, community and philanthropical organisations (Inaka 2020). The significant input of international financial institutions was taken into consideration in crafting both pieces of legislation. President Joseph Kabila restored the DRC's relationship with the International Monetary Fund (IMF) and World Bank after he became president (Trefon 2010). The financial institutions coached, nudged and helped Kabila's government conceptualise and enact the Investment Code, the Labour Code as well as the Mining

Code of July 2002, the Forestry Code of August 2002, and the Code of Ethics and Good Conduct of November 2002 (Mazalto 2009).

Personal interviews with a Member of Parliament (MP), assistant managers in the Labour Department and a sociologist respectively confirmed that these institutions influenced the outcomes of labour market legislation and, therefore, its reconstruction. This raises serious questions of legitimacy. The MP critiqued the pressure of the international financial institutions on the DRC thus:

> All these — France, Belgium, the United States — want to extract our diamonds, cobalt, coltan, timbers, etcetera. So, they sent that World Bank to oblige us to enact the Mining Code and the Investment Code following their injunctions. (Member of Parliament, Personal Interview, Kinshasa, June 2017)

Because this legislation was passed before the transitional government came into power, there are questions, however, as to its legality. As two Sociologists pointed out, the MPs who crafted the legislation at the time had been co-opted into these positions or had been appointed directly by Laurent-Désiré Kabila in July 2000. The drafting of the legislation in 2002 was thus based on a marked imbalance of power between the new president, Joseph Kabila, and legislators. One of the sociologists explained it this way:

> I am not against the 2002 Code of Investment... But my concern is with its legislators at that time. Who voted for these Members of Parliament? Did they stand for the interests of people? ... Kabila just co-opted them. So that parliament of the appointed MPs was just a rubber-stamping machine for the Kabila family. (Focus Group Discussion, Kinshasa, March 2017)

This seems to suggest that local actors could not resist the views of the international financial institutions in the drafting of the two codes. What is also crucial is that the reforms of the public sector labour market assumed a different form, linked to the fact that they were only tackled during, and not before, the transitional period.

The impacts of the Pretoria Agreement on the public sector were rather inevitable. The public administration in the DRC is the backbone of government and the primary employer in the formal labour market. Therefore, with the transitional government it became imperative to reform the administration as a basis of improving the public labour market (Kalulu 2013).

With the influence of international partners, the government went about conducting the reforms by creating an office responsible for the formulation and implementation of these reforms. It passed Decree No. 2003/035 of 13 November 2003 on the *Commission interministérielle de pilotage de la réforme de l'administration publique* (Interdepartmental Steering Committee for the Reform of the Public Administration, CIPRAP) (Diumasumbu 2008). CIPRAP was mandated to devise strategies for resolving a long list of issues in the public administration like the increasing levels of politicisation, clientelism and corruption; late and underpayment of salaries; the lack of equipment; absenteeism and forms of parallel employment (ghost workers); poor communication systems; and the privatisation of state assets (Moshonas 2014). The United Nations Development Programme (UNDP), the Belgian Technical Cooperation, IMF, World Bank, and French and Canadian organisations provided assistance to CIPRAD for the reforms (Diumasumbu 2008). The reforms sought a revision of the legal framework, census of the civil service, audit of payroll procedures, restructure ministries, establish a retirement programme and improve human resources management (Moshonas 2014).

Implementation of the reforms started in 2005. But they all failed or were abandoned for three reasons. First, the reforms were largely designed abroad and did not suit the environment they were implemented. The reform architecture underestimated the DRC's complex political and cultural environment, one that lacked good governance and accountability and high levels of predation on state resources (Trefon 2010). This is why the reforms failed, despite the influence and participation of many international actors (Mazalto 2009). Second, the census of public servants was to be done through a biometrical registration of employees' fingerprints (IMF 2007). This failed too. Though the census was conducted, it proved unreliable because of irregular employees on ministers' lists as well as ghost workers that government inspectors and agents who were implementing the census deceitfully included. The census validated irregularities instead of correcting them (Moshonas 2014). Third, the labour reforms failed because of lack of political will. Ministers blocked the restructuring of ministries in protest against labour reforms that supported the Ministry of Civil Service as the sole recruiter of civil servants (Moshonas 2014). The ministers sought to maintain the disorder in the public service—and benefit from it (Trefon 2011).

A former Minister identified how the political elites in the power-sharing regime obstructed the reform of the public service for self-interest:

> We knew that there was an outcry from the workers because of the delays in the reform and the failure to deliver the Mbudi Agreement. … We were under fire from the media, trade unions, the Catholic Church, and the donors. … [But] we couldn't waste [renege on] our advantages. Our comrades were waiting for their appointments, and the reform would have prevented them from finding jobs. (Former minister, Personal Interview, Kinshasa, July 2017)

The proposed reforms directly challenged the prerogatives legally granted to the parties of the transitional government. Consequently, these political elites artfully acted in opposing and blocking any reduction of these prerogatives.

It is plausible that these labour market reforms were conceived by international actors and imposed on political authorities in the DRC (Mazalto 2009). The case of the Mbudi Agreement questions this argument, however, by showing how local elites conceived and attempted to implement own-grown labour reforms. Therefore, the authorities and international actors should share responsibility for the failure of public administration reform policies (Diumasumbu 2008). More interestingly, the power-sharing agreement caused significant contradicted between the principles of the power-sharing structure and existing labour legislation. This contradiction was the underlying factor behind politicians disrupting the implementation of the reforms. From a legal perspective, the political context of the transition did not favour the implementation of the reforms. As the transitional constitution prevailed over other laws, any labour reform was supposed to be in line with the constitutional provisions of power sharing. Therefore, any labour legislation reform would, by definition, reproduce or maintain the effects of power sharing on the reconstruction of the labour market.

Labour Reforms After the Transition (2007–2018)

The new national constitution announced in February 2006 and Joseph Kabila's assumption of the presidency in November 2006 officially marked the end of the power-sharing regime. This introduced a new political order in the DRC that facilitated the implementations of some labour market reforms in the public and private sectors.

The labour reforms initiated during the transition government continued into Joseph Kabili's first term in office (i.e., 2006–2011). However, they did not achieve expected outcomes in the public sector. By 2010 they were considered a failure and all donors had either disengaged or reduced support (Moshonas 2018). This is partly because Cabinet, headed by Prime Minister Adolph Muzito, did not wield real power to implement reforms while Kabila's inner circle focused on consolidating power (Trefon 2010).

During Joseph Kabila's second term in office (2011–2019), Matata Ponyo, a significant figure in Kabila's inner circle, was appointed Prime Minister (2012–2017). Matata Ponyo resumed the previous reform agenda but renamed it the Revised Civil Service Reform Strategy (Ministère de la Fonction Publique 2014). His cabinet tried to implement five key labour policies including, a census of all civil servants, establishment of a retirement programme, a wage payment policy called *la bancarisation*, a rationalised and computerised payroll management scheme, and the Status of Civil Servants Act, No. 16/013 of July 2016 (Moshonas 2018).

The revived reform sought to craft labour legislation for the public administration in line with the 2006 constitution (Ministère de la Fonction Publique 2014). It was supposed to pave the way for a new census of civil servants after all elderly personnel had retired. The statistics would help establish effective recruitment systems, make merit-based appointments and judicious promotions and a rational management of civil servants. However, reform was delayed by six factors.

First, the government did not start by passing Act No. 16/013 to guide the implementation of the reforms as planned but did this last. This procedural error resulted in both the 2006 Constitution and the Status of a Civil Servants Act No. 81/003 of 17 July 1981 regulating the employment of public employees. It bred conflicts between provincial governments and the national government. Many provincial authorities claimed that the 2006 Constitution gave them rights to manage public servants under the authority of provincial governments. However, the Civil Service Department insisted that it was the sole competent authority to regulate public servants as long as Act No. 81/003 was not explicitly replaced or repealed (Ministère de la Fonction Publique 2014).

Second, the retirement programme was poorly implemented because of unfair retirement procedures based on inaccurate statistics. IMF (2007) estimates of 2007 suggested that almost 100,000 public workers above the official retirement age of 55 continued being engaged in work because retirement benefits were insufficient.

Third, the 2013 biometrical census recorded the number of civil servants at 793,615. Of these, 485,196 (35%) were *Nouvelles Unités* (new employees not yet captured in the administrative system)[5] and 30,260 (4%) were ghost workers or 'individuals on the payroll who do not exist' (Ministère de la Fonction Publique 2014). Although the synchronisation of data from the biometrical census has been pending since 2014, state authorities proceeded to hire civil servants. In 2012 Prime Minister Ponyo banned all recruitment of civil servants until 2016, because of the high levels of uncontrolled recruitment (Ministère de la Fonction Publique 2017).

Fourth, *bancarisation*, the reform of the wage payment policy, both positively and negatively impacted the labour market. The reform meant to implement the payment of all civil servant salaries through the banks. As a locally driven project, its successful implementation in 2013 surprised international donors to the extent that the World Bank and the IMF decided to reactivate their support for that reform (Moshonas 2019).

Fifth, the payroll reform addressed problems with the payment of salaries in the public administration. From 2007 to 2010, the computerised payroll management system replaced the manual system and formed the basis for the operationalisation of *bancarisation*. However, the new system could not eradicate the opacity, malpractices and corruption in the payment of salaries (Moshonas 2019).

Sixth, Act No. 16/013 of 15 July 2016 on the Status of Civil Servants was passed to align the administration with the national constitution. The act introduced some important innovations by increasing the minimum

[5] *Nouvelles Unités* were new recruits who did not have a civil service registration number because they were not yet integrated into the public administration system. They were recognised only by their employers (line ministries, provincial governments, and public enterprises) (Ministère de la Fonction Publique 2014).

age for employment from 15 to 18, a distinction of status between national and provincial civil servants, and gender equality among civil servants (Inaka 2020).

International partners, the Civil Service Department, various government representatives, and some unionists declared the reform a failure on 15 July 2015 (Radio Okapi 2015b). Nevertheless, the government held the resolve to fix the issues around retirement and biometrical census (Moshonas 2019).

Empirical data suggests that the reform of the public administration had no significant impact on the labour market. It shows that the refusal to include provincial civil servants in the national database and payroll meant that they could not draw on certain rights linked to their status as civil servants. Because the labour market was bad, civil servants often only found employment in their middle ages, after a long period of unemployment.

> Some young civil servants were hired when they were already over 40 because of unemployment. They were forced to retire as they were 55 years old. They barely worked for 20 years. (Retired unionist, Personal interview, Kinshasa, December 2016)

Many considered Prime Minister Matata Ponyo's decision to freeze recruitment for five years as the main reason people could not get jobs in the civil service between 2012 and 2016. A financial manager used an estimate of the number of Congolese who graduated between 2012 and 2016 as proof that Matata Ponyo's decision was an error of judgement:

> Imagine that all our universities produce more or less 100,000 graduates per year. In four years, they could be 400,000 graduates. ... These 400,000 persons couldn't find jobs in state services. For me, that decision was unworthy. (Financial Manager, KACA, Personal Interview, Kinshasa, November 2016)

Some informants, however, perceived *bancarisation* as having positive effects on both public and private labour markets. A manager of the Labour Department explained that *bancarisation* improved the system of paying wages and banking:

> Today, I receive my salary anywhere. With my bank card, I don't need to travel with a bag full of money even when I go abroad. (Likwalo, manager, Labour Department, Personal Interview, Kinshasa, February 2017)

A manager at the National Agency for the Promotion of Investments (ANAPI) explained how the *bancarisation* reform had domino effects on the private labour market and that it, for example, helped in creating job opportunities in the banking, transport, and private security sectors. Private banks opened up new branches in provinces to get closer to clients. Companies that transported money for banks also found opportunities to expand their business and hire new employees. New branches also meant the appointment of private security companies (Gbadolite, manager, ANAPI, Kinshasa, March 2017).

Act No. 16/013 of 15 July 2016 on the Status of Civil Servants resolved two key issues. These included the inequality in status between national and provincial public servants, and the contradictions between constitutional provisions and existing labour legislation. Implementing the act granted the same rights and privileges to public servants in the provinces as those enjoyed at the national level. President Joseph Kabila's second term in office thus saw the relaunch of the reform process.

In the private sector, the labour reforms included the enactment of two important acts meant to regulate the labour market of the sector after the transition. First, Act No. 16/010 of 15 July 2016 modified and extended Act No. 2002/015 of 16 October 2002 or the Labour Code (République démocratique du Congo 2016). It resolved outstanding issues relating to child labour, discrimination against married women and the lack of legal protection for employees with HIV/AIDS. The 2002 Labour Code had contravened ILO's Convention No. 182 (ILO 1999) which stipulates that 'the term child shall apply to all persons under the age of 18' (Dennis 1999). The 2002 Labour Code had also upheld the colonial legal requirement for married women to show written permission of husbands before being hired. Many employers also chose to misinterpret the 2002 Labour Code on discrimination against people living with HIV/AIDS (Yav & Associates 2017).

The situation on the ground suggests that it would be difficult to effectively implement these new laws. A human rights activist reflected on his experience with street children and suggested that the new legal position against the employment of under-18s was ineffective and would not help the children:

> Many kids work in the street because of poverty. They are accused of witchcraft. So, does the government have the means of taking care of these kids who have no support? … I think the act won't bring changes. (Human rights activist, Personal Interview, Kinshasa, March 2017)

An economist stressed the linkages between poor families and child labour, arguing that the act cannot be enforced at family level:

> Many kids work for their families in many rural areas. … I don't see how the act will deal with these issues. (Economist, Personal Interview, Free University of Kinshasa, April 2017)

On the discrimination against married women, a journalist considered the scarcity of formal job opportunities as the main obstacle to the effective enforcement of the new legal dispensation:

> Being a married household woman is almost *passé*. . . . As men say it nowadays: 'If I pay for [child]watcher, my wife pays for the electricity'. . . . For me, the problem is job opportunities are rare for men, there is no hope for the end of that injustice against married women. (Female journalist, Personal Interview, Kinshasa, March 2017)

These comments show that the new legal provisions against child labour can easily be enforced in the formal labour market. However, most kids working on the street or in rural areas are part of the informal sector. It thus becomes clear that practices of using workers who are under 18 are not fully addressed by the 2002 Labour Code.

One of the unexpected topics arising in this regard is the enactment of Labour Act No. 17/001 of 8 February 2017 on the creation and organisation of outsourcing of labour functions in the private sector (République démocratique du Congo 2017). The act was unexpected, considering that outsourcing had already been widely critiqued for abuse of workers. Some observers note that the legalisation of outsourcing undermines labour legislation that is extremely protective of the national workforce in the DRC (van Impe 2012). In fact, this is a moot point: the legislation is protective only in theory because workers were historically subjected to labour exploitation and unfair labour practices since colonisation. The new law grants companies the legal means to justify their practices, while the national workforce seems exposed to labour exploitation.

It is noteworthy that while private labour market reforms stalled during Prime Minister Muzito's term (2008–2011), they were more successful under Prime Minister Matata Ponyo because of his close affiliation with Kabila's inner circle. Political and the power relations among politicians prolonged implementation of labour market reforms.

Changes and Challenges in the DRC's Post-war Labour Market

Problems in post-war labour market reforms affect the functioning of the market. Statistics on the labour force, employment, renumeration and informal employment from official labour surveys, the informal sector and household consumption (*Enquête 1, 2, 3 sur l'emploi, le secteur informel et sur la consummation des ménages* in French), show trends that indicate changes and the challenges in the labour market. The surveys conducted in 2004 and 2012, respectively, are informative in this regard. The data from the surveys show a general trend of declining levels of employment which suggest that conditions have not improved since the end of the 1998–2002 war. The data actually also suggests increasing unemployment and underemployment as well as the evolution, role and impact of informal employment on the post-war labour market.

Economically Active Population and Employment

A correlation exists between the economically active population (workforce) and the employment rate in the DRC. The increase in the workforce since the end of colonial rule owes to the population growth of about 2.9% a year (Herderschee et al. 2011). But the employment rate has been trending in the opposite direction. In 1955, Bezy (1957) indicates that there were 1,000,000 employees—which means 40% of the workforce—which increased to 41% in the 1960s and 42% in the 1970s (Manwana 1987). Since the late 1970s, levels of employment have declined while the workforce increased rapidly. Levels of employment have declined because of economic regress and socio-political crises and wars. This trend has not changed in the post-war period. Data from the 2012 Survey shows that the workforce in the DRC represented 56% of the total population or 33.6 million people aged between 15 and 64 in 2012 (INS 2014). Among them, 27.7 million were either formally or informally employed. The workforce was also skewed in favour of men, who constituted 58.8% compared to 53.2% women. These figures in urban areas were 51 and 37.5% for men and women, respectively. Though the working-age population (15–64 years old) grew from 27.6 million to over 40 million between 2005 and 2012, workforce participation in the labour market declined from 74 to 67%, or from 26.4 million to 25.6 million (Aterido et al. 2017).

Though the foregoing statistics provide the most basic information on the size and structure of the DRC workforce, they do not give a full picture of the relationship between workforce growth and challenges in the post-war labour market. They are less appropriate for fully assessing the impact of government labour market policies on the DRC employment situation. For this it is necessary to turn to the statistics on employment.

The 2012 survey identified five main sectors of employment in the DRC including, public administration, parastatals, the formal private sector, the informal non-agricultural sector and the informal agricultural sector (INS 2014). It found that the total number of employees had increased from about 19 million (21.8%) in 2005 to over 27.7 million (28.6%) in 2012. Agriculture employed the largest proportion of the country's labour force (71.2%), followed by trade or services (24.4%), and industry (4.4%). The agricultural sector therefore employed more than 16 million people and more than 90% of the rural workforce. The informal sector employed 88.6% of the labour force in 2012 and was, therefore, a more important source of employment than the formal sector, which employed only 11.3% of workers and of which 8.7% were in the public administration and 2.8% in the formal private sector. Many formal employees were university graduates and averaged 42.9 years old in the public sector and 31.7 years old in the formal private sector. Of the employees in the public administration, 86% had employment contracts and 71% received pay slips, compared with 60 and 68%, respectively, in the private sector.

The 2012 Survey indicated that average incomes varied by sector and the employee's socio-professional status, urban or rural residence and gender. Employees in services, commerce and industry earned more than those in agriculture. In the public administration, employees of large public enterprises (2.3% of employees) were better paid than other civil servants. Yet employees in the formal private sector were even better paid, and managers in the formal private sector and executives in the formal public and private sectors earned the highest incomes. In general, workers in urban areas earned more than those in rural areas. The proportion of employees who earned less than the minimum wage was 44% in Kinshasa, 66% in other urban areas and 88% in rural areas. Women earned less than men. Overall, over 85% of workers were so poorly paid that they were unable to make ends meet and lived in a state of poverty.

These statistics are highly debated and subject to varying interpretations. One problem is the question of the reliability of any statistics issues by state institutions and leaders of the ruling party. For instance, Prime Minister Matata Ponyo used media and state publications to claim that his leadership and improvements in state institutions led to economic growth from 2008 to 2016 and the creation of over 50,000 formal jobs per year (Radio Okapi 2015a; Matata Ponyo 2016). Many politicians rebutted Ponyo's arguments and accused him of overestimating the socioeconomic impacts of the economic growth that was achieved during his period in office. Among them, the governor of the Central Bank of the Congo, Deogratias Mutombo, asserted that there had not been any economic growth during this period because of a fall in the price of cobalt and copper, a weak business environment in the country, and the lack of strategies for generating more income. The *Fédération des Entreprises du Congo* (Congolese Business Federation) president, Albert Yuma, argued that the unhealthy inflation rate at the time, depreciation of the local currency, and over-taxation of businesses were not indications of economic growth (Bouvier and Omasombo 2017). Adolph Muzito, who had preceded Matata Ponyo as Prime Minister (2008–2012), refuted Ponyo's arguments and accused the latter's cabinet of falsifying statistics by classifying informal self-employees as formal (Muzito 2015). Two studies combined the data from the 2012 Survey and the World Bank to determine that the DRC's economic growth under Ponyo did not in fact lead to any increase in the rate of employment (Aterido et al. 2017; Sumata 2020). Sumata (2020: 2) proved how the growth in the economy from 6.23% in 2008 to 9.47% in 2015 was inversely mirrored by an increase in the unemployment rate from 5.55 to 7.84%, respectively.

The data on the workforce and employment indeed show few positive changes in the labour market after the 1998–2002 war. As a result, the average annual growth rate of 5.3% of the country's workforce (aged 15–64) represents a challenge because new entrants to the labour market scarcely find jobs. Although employment and workforce growth evolve concurrently, the labour market fails to absorb many of the young people who join the workforce. The high level of underpayment suggests that the state has struggled to implement new wage policies since 2003. The absence of employment contracts and pay slips offer robust evidence of poor employer-employee labour relations, job insecurity and the violation of the labour legislation, which stipulates that employees must be provided with written employment contracts. As a result of this malpractice, trade unions have found it difficult to defend the rights of employees who are not issued with employment contracts.

Overall, all features of the workforce and employment are shaped by neopatrimonialism, poor governance, clientelism and weak institutions, which are explained by theories on the African state and global labour history.

Unemployment and Underemployment

Although it was infrequent during the colonial period (as it only occurred during the Great Depression of the 1930s), unemployment is a long-standing reality in the DRC (Bezy 1957; Jewsiewicki 1977). The 2012 survey indicated that the unemployment rate increased from 1.9% in 2005 to 2.2% in 2012 (INS 2014). This rate was shaped by sex, age, education, urban or rural residency, length of unemployment and types of job-searching methods employed. Unemployment affected men more than women, at 3.6 and 5.2% respectively, regardless of rural or urban setting. The period 2005–2012 was marked by a significant increase in the working-age population and in unemployment with more than 40% of the unemployed being young adults. Aterido et al. (2017) emphasise that the DRC is marked by an uncommon level of highly educated people experiencing unemployment. Unemployment in the DRC is also a predominantly urban phenomenon as 83% of the unemployed live in urban areas. Urban men accounted for 50.9% of the unemployed in 2012, with rural men accounting only for 11.1% (Aterido et al. 2017). Unemployment is also characterised by increases in the number of long-term unemployed persons. From 2005 to 2012, for example, the length of unemployment and job seeking increased from an average of over five years to over eight years (Kankwanda et al. 2014). The average length of unemployment was also higher among *des primo-demandeurs* (first time job seekers, at 98.4 months) than among *des anciens occupés* (the formerly employed, at 97.2 months). The 2012 survey suggested that the majority of unemployed people (74.9%) drew on personal relationships with family and friends when looking for employment, with 15% seeking employment directly from employers and about 5.4% reacting to job advertisements.

Unlike the 2004 and 2012 surveys, studies that included data from the demand side of the labour market found additional causes of unemployment in the DRC. These included hiring requirements, the types of employers and the poor business climate. Fresh graduates and long-term jobseekers often fail to meet hiring requirements like professional experience, language skills and outstanding labour skills. The limitation of job creation due to the weak business climate closed the window of opportunity for many jobseekers (Aterido et al. 2017; Sumata 2020).

The overall underemployment rate in the DRC reached 72% of the workforce in 2012. Underemployment was more visible in rural areas and among the youth: 71% of rural workers were classified as underemployed as against 47% in urban environments. In addition, Kankwanda et al. (2014) show that underemployment also tended to go together with low pay and labour exploitation, abuse and insecurity. In formal respects, 2012 survey data on working hours in relation to underemployment are respectable since they give an idea of the linkages between underemployment and underpayment, poverty and labour precarity. Empirical data from the fieldwork conducted for this study shows in addition that underemployment is of greater concern to workers who have experienced job downgrading and labour flexibility.

Overall, the effects of unemployment in the last three decades affect the labour market. That being said, it is important to reiterate that the persistence of neopatrimonialism, with its corollaries of social injustice and inequality, corruption and poor governance, have not made it easier for the authorities to better implement labour policies aimed at reducing unemployment. Thus, it must be argued that the efforts by the government against underemployment and unemployment have not yet achieved the expected results since the 1998–2002 war. Since the reduction and regulation of unemployment and underemployment are still major issues in the DRC, informality appears to be a way of survival for many unemployed, underemployed and permanent but underpaid workers.

Informal Employment

Informal employment shapes the economic life of many Congolese and their basic livelihoods since the colonial era, even as the state has sought to eradicate, control, or formalise this informality. Although there was a relatively large formal labour market in the Belgian Congo, especially among white residents, informal employment was numerically more important, even in the urban areas. In 1955, towards the end of the colonial period, 39% of the urban workforce was in the formal sector and 61% in the informal sector (De Herdt and Marysse 1996). By 1961, the number of informally employed had risen to 70.9% (De Herdt and Marysse 1996) in the context of the post-independence secessions, rebellions and political-bureaucratic conflicts (Ishako 2018). In response, the government often brutally destroyed the working sites of informal workers or relegated some of them to the rural areas (Ishako 2018).

In the aftermath of *Zaïrianisation* and the Mobutu regime's poor governance in the late 1970s, the growth of informal employment reached its high peak in the 1980s, with 80% of workers being employed informally (MacGaffey 1983). With worsening socioeconomic conditions in the early 1990s, side effects of the looting and mismanagement that was taking place, the informal sector rose to employ 95% of all employed individuals (De Herdt and Marysse 1996).

The Mobutu regime had limited options to control or formalise the informal sector. Some informal self-employed were officially registered and required to pay site-fees in specific locations (Ishako 2018). Interestingly, over time some illegal jobs shifted into the informal and then the formal sector. De Herdt and Marysse (1996) show, for example, how women who engaged in illicit currency exchange contributed to the informalisation and, then, formalisation of their activities in the 1990s.

After the 1998–2002 war, the 2012 Survey shows that 88.6% of all workers were employed in the informal sector (INS 2014). This informal employment is boosted by commercial activities and by women (55%) who held more than half of all informal jobs. It was characterised by poor labour relations and conditions, with over 96% of employees not having written contracts, hardly any receiving salaries and legal standards governing working hours not applying. More than half (56.4%) of all informal enterprises in the Congolese agglomerations exercised their activities at home.

As before the 1998–2002 war, the government continued to use repression, control, or formalisation afterwards in an attempt to regulate informal employment. The provincial government of Kinshasa, for example, tried a succession of campaigns—*Operation coup de poing*

(Operation crackdown) in October 2004, *Operation Kin propre* (Operation clean Kinshasa) in 2007, and *Operation Kin Bopeto* (Operation smart Kinshasa) in 2019—to clean the city from informal workers who set up kiosks, street garages, restaurants, hair salons, vending tables and pubs without respect for town planning rules, thus increasing the town's insalubrity. Despite police brutality employed during these campaigns, informal workers subsequently re-emerged with their businesses.

Informal employment in the DRC is, therefore, marked by a complementarity between the 'autonomous informal sector' and the 'informal sector as an integrated and subordinated part of the same economy'. The autonomous informal sector occurs when people resort to informal employment to meet their basic needs. However, even though the informally employed are not recognised by law and face labour market insecurity and precarity, they have been participating in the integrated and subordinated parts of the economy for ages.

Further, the state's temptation to regulate informal employment shows how that sector is both a necessity for sustainable livelihoods and an obstacle for a 'normal' functioning labour market. There is no need to remind us that the government long ago recognised the importance of informal employment as a necessity for the survival of its citizens. Thus, it has been trying to regulate it. Moreover, considering the statistics cited above on unemployment, it can be argued that informal employment has been playing a major role in job creation, income generation and production, before, during, and after the war of 1998–2002. It enables many people to survive economic crises, supplement their livelihood and provide means to avoid proletarianization. Besides, the impact of informality is vividly observable in the labour market because of the informal market's absorption capacity and contributions to people's livelihoods.

It is thus appropriate to argue that although informal employment has been allowing many people to survive, it also poses a challenge to the improvement of the post-war labour market. These realities and challenges linked to informal employment in the DRC can be captured by theories of the African state, which highlight how bad public governance, with neopatrimonialism and weak institutions, continues to disrupt economies and labour markets.

Conclusion

This chapter showed the main aspects that shaped the labour market throughout the history of the DRC. A continuing feature has been the presence of forms of coercion over the labour force. Whereas in the Congo Free State this was forced labour approximating slavery, it became coerced labour during the Belgian Congo, and then unfree since independence because of forms of control exerted over trade unions and the flouting of labour laws protecting the rights of workers. The second continuing feature is the influence of the international system on the local environment. Global colonialism and empire building created the first exploitative labour systems in the DRC. The Cold War then saw Mobutu align himself with the USA and win its support, allowing him to rein as dictator for his own benefit and that of a small coterie of friends and relatives. The fall of the Eastern Bloc and accompanying decline of the USA's support to former allies saw the collapse of Mobutu's regime. This created room for a range of local and regional geopolitical interests to make claims on the DRC. Due to the lack of a unitary or bipolar international political force, multiple political interests began to exert power over the country, creating parallel labour markets in competition with each other.

The chapter also presented factors that conditioned and inhibited attempts by the state to achieve post-war labour market reconstruction between 2003 and 2019. It demonstrated that the state undertook measures that insufficiently facilitated labour market reconstruction in the DRC. In fact, the power-sharing deal had a negative impact on the reconstruction of the post-war labour market during the transition period (2003–2006). The power-sharing provision legally politicised the labour market and thus hindered its reconstruction. International efforts to help the DRC reform did not acknowledge the complexities of the political, cultural, economic and administrative realities.

The election of a new government and adoption of a new constitution in 2006 created a conducive environment for labour market reforms in the public and private sectors. But another period of hiatus ensued from 2007 to 2011, when reforms by the government of Prime Minister Muzito were blocked and undermined to protect private interests by President Joseph Kabila's inner cabal. The situation changed under Prime Minister Matata Ponyo who implemented a series of labour market reforms. These were only partially successful, however. Whereas the 2013 policy of *bancarisation*, the new Labour Code of 2016, and the new act on the status of civil servants somehow improved the employment conditions of public workers. The implementation of the reforms was marked by significant delays, however, which happened at the expense of the labour force: the 2016 Act on the Status of Civil Servants. These delays demonstrate that the government was very inefficient in implementing processes to improve the labour market.

The chapter also analysed challenges that the labour market in the DRC faces in the process of reconstruction. Using statistical data on the economically active population, employment, unemployment, underemployment, and informal employment, the chapter supports the argument which states that the labour policies of the state insufficiently facilitated the labour market's reconstruction in the post-war period. The first challenge is that the rapid increase in the workforce inversely coincided with the extreme depletion of opportunities in the labour market. This resulted in the increase of unemployed and inactive persons. The second challenge is the poor employment situation in the labour market. While levels of employment do not grow, insecure employment, underpayment and underemployment persist without efficient labour protection.

The persistently high unemployment rate and the rise of underemployment are also among the main challenges that impact negatively on the labour market. While unemployment affects the vast majority of the workforce, forcing them to survive in pervasive poverty and anxiety, marked by social stigma, underemployment strengthens the casualisation of jobs and the precariousness of the lives of the underemployed.

Given the consequences of poor labour conditions and relations, unemployment and underemployment, using informal employment as an avenue for alternative possibilities has allowed many informally employed people to survive for over a century. Despite the important role in many workers' lives, informal employment contributes to a vicious cycle of inequality and informality, which in turn impacts on the improvement of the labour market.

Overall, the state's labour market policies did not prove conducive to the regulation of informal employment, the creation of *secure* jobs, the amelioration of job-searching methods and the improvement of the lives of the unemployed. These challenges of the labour market are also reflected in labour market institutions, to which we turn in the next chapter.

References

Abbeloos, Jan-Frederick. 2008. "Belgium's Expensionist History Between 1870 and 1930: Imperialism and the Globalisation of Belgian Business." in *Europe and Its Empire*, edited by Csaba Levai. Pisa: Plus.

Ahmad, Gulzar, and Muhamad Safeer Awan. 2017. "Colonialism and Its Socio-Politico and Economic Impact: A Case Study of the Colonized Congo." *The Dialogue* 12(3):311–20.

Aterido, Reyes, Alvaro Gonzalez, Dino Merotto, Carly Petracco, and Javier Sanchez-Reaza. 2017. *Democratic Republic of Congo Jobs Diagnostic*. Washington, DC: World Bank.

Au, Sokhieng. 2017. "Medical Orders: Catholic and Protestant Missionary Medicine in the Belgian Congo 1880–1940." *BMGN-Low Countries Historical Review* 132(1).

Autesserre, Severine. 2010. *The Trouble with the Congo: Local Violence and the Failure of International Peacebuilding*. New York: Cambridge University Press.

Bach, Daniel C. 2011. "Patrimonialism and Neopatrimonialism: Comparative Trajectories and Readings." *Commonwealth & Comparative Politics* 49(3):275–94.

Basila, Jean-Pierre Kilenda Kakengi. 2004. *L'affaire des 315 magistrats de Kinshasa: Une purge néo mobutiste*. Paris: Editions L'Harmattan.

Bezy, Ferdinand 1957. *Problemes Structurels de l'economie Congolaise*. Louvain: Nauwelearts.

Bossekota, Leopold Watshia, and Jacques Kiseta Sabiti. 2001. "RD Congo: Dernier pays du Monde ?" in *Rapport d'OBSAM*, edited by Leopold Watshia Bossekota and Jacques Kiseta Sabiti. Kinshasa: Universite Technologique Bel Campus.

Bouvier, Paule, and Jean Tshonda Omasombo. 2017. "Conjonctures Congolaises 2016. Glissement politique, recul économique." Pp. 13 in *RDC 2016: une année électorale sans élections*, edited by Aymar Nyenyezi Bisoka, Sara Geenen, An Ansoms, and Jean Omasombo Tshonda. Paris: Éditions L'Harmattan.

Brausch, Georges. 1961. *Belgian Administration in the Congo*. Oxford: Oxford University Press.

Congo Belge. 1922a. "Arrêté Royal du 05 Fevrier 1922 portant Statut des fonctionnaires et agents du la Colonie, autres que les agents de l'ordre judiciaire." Bruxelles Bulletin Officiel du Congo Belge.

Congo Belge. 1922b. "Arrêté royal du 27 fevrier 1922 portant Statut du personnels Africains de la Colonie." Bruxelles Bulletin Official du Congo Belge.

Congo Belge. 1922c. "contrat de travail entre indigènes et maîtres civilisés." Bruxelles Bulletin officiel du Congo belge.

Congo Belge. 1922d. "Décret royal du 13 Mars 1922 portant le louage de services privés des Européens." Bruxelles Bulletin officiel du Congo belge.

Daniels, John. 1908. "The Congo Question and the "Belgian Solution"." *The North American Review* 188(637):891–902.

De Boeck, Filip 2004. "On Being Shege in Kinshasa: Children, the Occult and the Street." Pp. 155–73 in *Reinventing Order in the Congo: How People Respond to State Failure in Kinshasa*, edited by Theodore Trefon. London and New York: Zed Books.

De Clerck, Louis. 2006. "L'administration coloniale belge sur le terrain au Congo (1908–1960) et au Ruanda-Urundi (1925–1962)." *Annuaire d'Histoire administrative européenne* 18:187–210.

De Herdt, Tom, and Stefaan Marysse. 1996. *L'économie informelle au Zaïre.(Sur) vie et pauvreté dans la période de transition*. Paris: L'Harmattan.

Dennis, Michael J. 1999. "The ILO Convention on the Worst Forms of Child Labor." *American Journal of International Law* 93(4):943–48.

Diumasumbu, Raphaël Mukanga. 2008. "Réforme administrative de la République démocratique du Congo: leçons et travers du passé." *Afrique contemporaine* (3):95–106.

Ewans, Martin. 2003. "Belgium and the Colonial Experience." *Journal of Contemporary European Studies* 11(2):167–80.

Foucault, Michel 1975. *Surveiller et punir*. Paris: Gallimard.

Gondola, Ch. 1997. "Unies pour le meilleur et pour le pire. Femmes africaines et villes coloniales: une histoire du métissage." *Clio. Femmes, Genre, Histoire* 6(2):87–104.

Haskin, Jeanne M. 2005. *The Tragic State of the Congo: From Decolonization to Dictatorship*. New York: Algora Publishing.

Herderschee, Johannes, Kai-Alexander Kaiser, and Daniel Mukoko Samba. 2011. *Resilience of an African Giant: Boosting Growth and Development in the Democratic Republic of Congo*. Washington, DC: The World Bank.

Hochschild, Adam. 1998. *King Leopold's Ghost: A Story of Greed, Terror, and Heroism in Colonial Africa*. Boston and New York First Mariner Books Edition.

Houben, Vincent, and Julia Seibert. 2013. "(Un) freedom: Colonial Labor Relations in Belgian Congo and the Netherlands Indies Compared." Pp. 198–212 in *Colonial Exploitation and Economic Development. The Belgian Congo and the Netherlands Indies Compared*, edited by Ewout Frankema and Frans Buelens. New York and London: Routledge.

ILO, International Labour Organization. 1999. *The ILO Convention No. 182 concerning the Worst Forms of Child Labour Convention* [Online]. Geneva: The ILO. Available: http://www.ilo.org/dyn/normlex/en/f?p=NORMLEXPUB:12100:0::NO:12100:P12100_ILO_CODE:C182 [Accessed 22sd April 2018].

Ilofo, Ise Etenteke-Elongo. 1973. "Essai sur la Reglementation du Service National de l'Emploi au Zaire (Titre XI, Chapitre de l'ordonnance-loi No 67/310 du 9 aout 1967 portant code du travail)." in *Faculte de Droit*. Kinshasa: Universite Nationale du Zaire, Campus de Kinshasa.

IMF, International Monetary Fund. 2007. "Democratic Republic of the Congo: Selected Issues and Statistical Appendix." in *IMF country report No. 05/373*. Washington, DC: International Monetary Fund.

Inaka, Saint José 2020. "Post-war Labour Market Reconstruction: The Case of the Democratic Republic of the Congo." in *Doctoral dissertation, Sociology*. Pretoria: University of Pretoria.

INS, Institut National de la Statistique. 2014. "Rapport Global, Enquete 1–2–3. Resultats de l'enquete sur l'emploi, le secteur informel et sur la consommation des menages." in *Rapport Global* Kinshasa INS.

Inter-Congolese Dialogue. 2002. *Global and All-Inclusive Agreement on the Transition in the Democratic Republic of the Congo*. Pretoria.

Ishako, Peru. 2018. " La logique propre dans des activités informelles. Cas des vendeurs du Marché Gambela." in *Sociologie*. Universite de Kinshasa.

Jewsiewicki, B. 1977. "The Great Depression and the Making of the Colonial Economic System in the Belgian Congo." *African Economic History* (4):153–76.

Kairouz, Matthieu. 2016. "Ce jour-là: le 11 juillet 1960, le Katanga proclame son indépendance." in *Jeune Afrique*.

Kalulu, Théophile Yuma. 2013. "La Fonction publique congolaise (Zaïre) face aux défis de son renouvellement." Pp. 240 in *Les réformes du secteur public en République démocratique du Congo. Rétrospectives et perspectives*, edited by Camille Elatre Welepele and Hubert Lukunga Ntumba Cordosia Dakar.

Kankwanda, Grégoire, Timothée Makabu Ma Nkenda, Björn Nilsson, François Roubaud, Constance Torelli, and Jean-Michel Wachsberger. 2014. *Le marché du travail en République Démocratique du Congo en 2012: Principaux résultats de la Phase 1 de l'Enquête 1-2-3*. Kinshasa et Paris INS et UMR DIAL.

Kawu, Théodore Kaminar Nsiy. 2013. "Enjeux politiques et administratifs de la privatisation des entreprises publiques en République démocratique du Congo (RDC)." Pp. 37–58 in *Les réformes du secteur public en République démocratique du Congo.*, edited by Camille Elatre Welepele and Hubert Lukunga Ntumba. Dakar: CODESRIA.

Kisangani, Emizet Francois, and Scott F. Bobb. 2010. *Historical Dictionary of the Republic Democratic of the Congo*. Lanham: The Scarecrow Press.

Kodi, Muzong. 2008. *Corruption and Governance in the DRC During the Transition Period (2003–2006)*. Pretoria: Institute for Security Studies.

Larmer, Miles, and Erik Kennes. 2014. "Rethinking the Katangese Secession." *The Journal of Imperial and Commonwealth History* 42(4):741–61.

Leslie, Winsome J. 1993. *Zaire. Continuity and Political Change in an Oppressive State*. Westview Press.
Lumbu, Manasse Mangbanga. 2004. "Les Ministres de Mobutu (1965–1990), Memoire de Licence en Histoire." in *Department de l'Histoire, Faculte de Lettres et Sciences Humaines*. Kinshasa: Universite de Kinshasa.
Lwamba, Bilonda. 1985. "Histoire du Mouvement Ouvrier au Congo Belge (1994–1960): Cas de la Province de Kattanga, These de doctorat en Histoire." in *Faculte de Lettres, Departement de l'histoire*. Lubumbashi: Universite de Lubumbashi.
MacGaffey, Janet. 1983. "How to Survive and Become Rich Amidst Devastation: The Second Economy in Zaire." *African Affairs* 82(328):351–66.
Manwana, Mungongo. 1987. "Evolution de l'Emploi au Zaire: Avant et Apres l'Independance." *Zaire-Afrique* 220:595–615.
Marriage, Zoe. 2011. "Divisive 'Commonality': State and Insecurity in the Democratic Republic of Congo." *Third World Quarterly* 32(10):1891–1910.
Matata Ponyo, Mapon. 2016. *Qualite des institutions et resultats economiques en Republique democratique du Congo 1980-2015*. Paris: Editions L'L'Harmattan.
Matenda, Athanase Kyelu. 2002. *Fédération des Entreprises du Congo* [Online]. CBC. Available: www.winne.com/congo/to05.html [Accessed 9 May 2019].
Mazalto, Marie. 2009. "Governance, Human Rights and Mining in the Democratic Republic of Congo." Pp. 187–242 in *Mining in Africa: Regulation and Development*, edited by Bonnie Campbell. New York: Pluto Press.
Ministère de la Fonction Publique. 2014. *Le Journal du Fonctionnaire No. 1.* Kinshasa Syrius Presse.
Ministère de la Fonction Publique. 2017. "Cadre Stratégique Général Révise de la Réforme et Modernisation de l'Administration Publique." edited by Ministère de la Fonction Publique. Kinshasa: Journal Officiel
Moshonas, Stylianos. 2014. "The Politics of Civil Service Reform in the Democratic Republic of Congo." *The Journal of Modern African Studies* 52(2):251–76.
Moshonas, Stylianos. 2018. *Power and Policy-Making in the DR Congo: The Politics of Human Resource Management and Payroll Reform.* Brussels: Universiteit Antwerpen, Institute of Development Policy (IOB).
Moshonas, Stylianos. 2019. "Reform of the Public Wage System in the DRC: The Système Intégré de Gestion des Ressources Humaines et de la Paie and Its Prospects." Pp. 26–50 in *Negotiating Public Services in the Congo: State, Society and Governance*, edited by Stylianos Moshonas, Tom De Herdt, and Kristof Titeca. London: Zed Book.
Mpinga, Kasenda, and David J. Gould. 1975. *Les Réformes Administratives Au Zaïre, 1972–1973.* Kinshasa: PUZ.
Muzito, Adolphe. 2015. *Le Chomage en RDC: Hier, aujourd'hui et demain.* [Online]. The Eastern Congo Tribune. Available: https://edrcrdf.wordpress.com/2015/06/19/le-chomage-en-rdc-hier-aujourdhui-et-demain-selon-adolphe-muzito/ [Accessed 19 Septmber 2020].

Mwabila, Malala. 1979. *Travail et Travailleurs au Zaire: Essai sur la Conscience Ouvriere du Proletariat Urbain de Lubumbashi*. Kinshasa: Presse Universitaire du Zaire.

Ndaywel, Isidore è Nziem. 1998. *Histoire générale du Congo. De l'héritage ancien à la République Démocratique*. Bruxelles: Duculot Afrique Editions.

Ngoy-Kangoy, H Kabungulu. 2006. *Parties and Political Transition in the Democratic Republic of Congo*. Johannesburg: EISA.

Nzongola-Ntalaja, Georges. 2002. *The Congo: From Leopold II to Kabila. A People's History*. London: Zed Book.

Nzongola-Ntalaja, Georges. 2004. *From Zaire to the Democratic Republic of the Congo*. Uppsala: Nordic Afrika Institutet.

Oppong, Joseph R. 2007. *Democratic Republic of the Congo*. New York: Chelsea House

Parkham, Thomas. 1992. *The Scamble for Africa: The White Man's Conquest of the Dark Contienent for 1976 and 1912*. London: Abacus.

Radio Okapi. 2015a. *RDC: près de 50 000 emplois ont été créés en 2015, selon Matata Ponyo* [Online]. Available: https://www.radiookapi.net/2016/02/03/actualite/politique/rdc-pres-de-50-000-emplois-ont-ete-crees-en-2015-selon-matata-ponyo [Accessed 29 September 2020].

Radio Okapi. 2015b. *RDC: résultats mitigés de la réforme de l'administration publique* [Online]. Available: https://www.radiookapi.net/politique/2015/07/10/rdc-resultats-mitiges-de-la-reforme-de-ladministration-publique/ [Accessed 11 June 2019].

République démocratique du Congo. 2002a. "Décret n° 2002/065 du 05 juin 2002 portant statuts, organisation et fonctionnement de l'agence nationale pour la promotion des investissements, en sigle «ANAPI»", edited by Cabinet de la President de la Republique. Kinshasa: Journal Officiel de la Republique Democratique du Congo.

République démocratique du Congo. 2002b. "Loi n°004 du 21 février 2002 portant Code des Investissements" in *6*, edited by Cabinet du President de la Republique. Kinshasa: Journal Officiel de la Republique Democratique du Congo.

République démocratique du Congo. 2002c. *Loi n°015/2002 portant Code de Travail*. Kinshasa: Journal Officiel de la Republique Democratic du Congo.

République démocratique du Congo. 2003. *Constitution de la transition de la République Démocratique du Congo*. Kinshasa: Journal Officiel.

République démocratique du Congo. 2004. *Contrat social de l'innovation entre le Gouvernement et les syndicats de l'Administration publique et interprofessionnels*. Kinshasa: le Gouvernement de transition de la République démocratique du Congo.

République démocratique du Congo. 2016. *Loi N° 16/010 DU 15 JuilletT 2016 Modifiant et Completant la loi N° 015-2002 portant Code du travail*. Kinshasa Journal Officiel.

République démocratique du Congo. 2017. *Loi No. 17/001 du 08 fevrier 2017 fixant les regles applicables a la sous-Traitance dans le secteur prive*. Kinshasa Journal Officel.

République du Zaïre. 1981. *Loi No. 81/003 du 17 Juillet 1981 portant statut du Personnel de Carriere des Services Public de l'Etat* Kinshasa: Journal Offciel.

Reybrouck, David Van. 2014. *Congo. The Epic History of a People*. New York: HarperCollins.

Reyntjens, Filip. 2009. *The Great African War. Congo and Regional Geopolitics. 1996–2006*. Cambridge: Cambridge University Press.

Soriano, Victor Fernandez. 2017. "Travail et Progres: Obligatory 'Educational' Labour in the Belgian Congo." *Journal of Contemporary History* 1(1):1–23.

Stearns, Jason K. 2012. *Dancing in the Glory of Monsters. The Collapsed of the Congo and the Great War of Africa*. New York: Public Affairs.

Sumata, Claude. 2020. *Emploi des jeunes et dynamique de l'entreprenariat en RD. Congo: Une evaluation des mecanismes d'auto-emploi*. Montréal: l'Université de Montréal.

Trapido, Joe. 2016. "Kinshasa's Theatre of Power." *New Left Review* 98:57–80.

Trefon, Theodore. 2010. "Administrative Obstacles to Reform in the Democratic Republic of Congo." *International Review of Administrative Sciences* 76(4):702–22.

Trefon, Theodore. 2011. *Congo Masquerade: The Political Culture of Aid Inefficiency and Reform Failure*. London: Zed Books Ltd.

Tshimpaka, Leon Mwamba, and Christopher Changwe Nshimbi. 2022. "Scaffolding the State: Faith-Based Organisations and Application of Democratic Principles in the DRC and Zambia." *Religion, State & Society* 50(1):76–95.

Tshombe, Lukamba-Muhiya 2017. "Public Administration in the Democratic Republic of Congo." Pp. 27–44 in *Public Administration in Africa. Performance and Challenges*, edited by Shikha Vyas-Doorgapersad, Lukamba-Muhiya Tshombe, and Ernest Peprah Ababio. Boca Raton: CRC Press.

Vaessen, Vincent. 2001. "Les Probelemes sur législations sur le contrat de travail de 1910 et 1922 au Congo Belge: deux intentions et deux modes de décisions opposés." *Revue belge de philologie et d'histoire* 79(4):1213–54.

van Impe, Andreas 2012. *Investing in the Democratic Republic of the Congo. A legal Guide*. Brussels: De Wolfs and Parterns.

Weisbord, Robert G. 2003. "The King, the Cardinal and the Pope: Leopold II's Genocide in the Congo and the Vatican." *Research Network in Genocide Studies* 5(1):35–45.

Williams, Eric. 2014. *Capitalism and Slavery*. New York: UNC Press Books

Yav & Associates. 2017. *Bref aperçu des innovations apportées au Code du Travail en République Démocratique du Congo* [Online]. Kinshasa: Yav & Associates Available: https://www.legavox.fr/blog/yav-associates/bref-apercu-innovations-apportees-code-22934.pdf [Accessed 19 March 2017].
Young, Crawford. 1965. "The Congo Rebellion." *Africa Report* 10(4):6.
Young, Crawford. 2015. *Politics in Congo: Decolonization and Independence.* London: Princeton University Press.
Young, Crawford, and Thomas Turner. 1985. *The Rise and Decline of the Zairian State.* Wisconsin and London: University of Wisconsin Press.

CHAPTER 5

Post-war Labour Market Reconstruction and Public Labour Market Institutions

This chapter considers the domestic context of labour markets in a nation-state. It continues the focus on the Democratic Republic of Congo (DRC) and analyses the contribution of public labour market institutions to post-war labour market reform and recovery. The chapter builds on previous chapters while considering historical, political, economic and social environments and conditions in the Southern African Development Community (SADC) region and the DRC. The labour markets in the SADC region including the DRC were unjust and characterised by conflict and war along with efforts to make the markets viable as they evolved. The chapter also draws on Peck's notion of the labour market as an arena of contestation, conflict, contradiction and complexity and argues that the contribution of labour market institutions to the remaking of the DRC's labour market between 2003 and 2018 was ineffective. The chapter focuses on the DRC and draws views from the literature, national policies and legislation and data gathered from interviews and focus group discussions (as detailed in Chapter 1 of this book). With this, the chapter highlights issues on several labour market institutions of the DRC. It shows that the Labour Department of the Government of the Democratic Republic of the Congo drew national policies and intervened in the labour market between 2003 and 2018 to solve various problems. The chapter also explains how key actors in the National Council of Labour (Conseil

S. J. Inaka et al., *The Reconstruction of Post-War Labour Markets in The Southern African Development Community*,
https://doi.org/10.1007/978-3-031-37357-2_5

National du Travail, CNT) were in conflict with each other and that this affected efforts to improve the labour market. The chapter further examines how the National Institute for Vocational Training (Institut National pour la Préparation Professionnelle, INPP) attempted to supply skilled workers to employers in the country and shows the serious problems the National Employment Office (Office National de l'Emploi, ONEM) faced in its work of linking employers and jobseekers. It further considers how the National Social Security Fund (Caisse Nationale de Sécurité Sociale, CNSS) attempted to implement new welfare strategies and reveals how the National Agency for the Promotion of Investments (Agence Nationale pour la Promotion des Investissements, ANAPI) struggled to attract investment and improve the business climate in the DRC.

Department of Employment, Labour and Social Welfare

The Department of Employment, Labour and Social Welfare, or the Labour Department, constitutes the national cabinet of ministers and three secretariats including, the General Secretariat of Social Welfare; the General Secretariat of Employment and Labour; and the General Inspectorate of Labour. The Labour Department also has three self-governing agencies namely, the National Employment Office; the National Institute of Vocational Training; and the National Social Security Fund

The Labour Department was created at independence in 1960 and was not spared from the deterioration of the country's public administration experienced from the time of Mobutu's regime up to the end of the 1998–2002 war. From 2003 onwards, the Labour Department attempted to implement four key labour market policies, which included measures against unfair and mass dismissal procedures (from October 2005); the Interprofessional Guaranteed Minimal Wage policy (from April 2008); the National Programme of Labour and Vocational Training (September 2008); and measures against unlawful practices of private job placement services (September 2008).

In October 2005, Minister Balamage issued two Ministerial Orders to deal with the issue of unfair dismissals. The orders were Ministerial Order No. 12/CAB.MIN/TPS/116/2005 on the modalities of dismissal of workers (Ministère de l'Emploi du Travail et de la Prévoyance Sociale 2005b) and Ministerial Order No. 12/CAB.MIN/TPS/117/2005, on warnings and notices prior to dismissing employees (Ministère de l'Emploi du Travail et de la Prévoyance Sociale 2005a). Ms Marie-Ange Lukiana, a new minister in the 2006 government, strengthened these instruments by issuing Ministerial Order No. 12/CAB.MIN/ETPS/038/08 of 8 August 2008 on the temporary prohibition of the mass dismissal of workers by labour inspectors (Ministère de l'Emploi du Travail et de la Prévoyance Sociale 2008a).

Although these orders seem to be achievements, the responses gathered from personal interviews for this book suggested that the situation concerning them was more complex. Contrary to the positive evaluation of a manager in the Labour Department, a trade union leader and members of *24 entreprises et subséquentes*, the unofficial pressure group of unfairly dismissed employees, found the instrument to be ineffective. A Manager in the Labour Department expressed the following view,

> Employers were firing their workers as they wanted. There was an amalgam of many politicians, our corrupted colleagues of the General Inspectorate who were umbrellas of many employers. ... The decisions were timely. (Likwalo, Labour Department Manager, Personal Interview, Kinshasa, February 2017)

In contrast, a unionist viewed the measures instituted by the Labour Department as ineffective:

> For them, it was good. But they did nothing to reintegrate fired workers in their jobs. Why didn't they take actions against these employers? (Alain, Senior Executive, CSGA,[1] Personal Interview, Kinshasa, December 2016)

The view of the trade unionist corroborates arguments made by members of *24 entreprises et subséquentes* who also considered the Labour Department ineffective since it did not enforce the decisions it had taken in favour of the dismissed employees. Composed of over 45,000 former employees from over 24 public entities and private companies who were victims of mass dismissals during the transition, the pressure group had been struggling to have the labour rights of its members restored since 2003. Although the Labour Department, the Supreme Court, Parliament,

[1] CSGA is the *Confédération Générale des Syndicats autonomes du Congo* (Congolese General Confederation of Autonomous Trade Unions).

the President and Prime Minister had ordered that the workers be reinstated, the Congolese Businesses Federation vetoed the implementation of the decision (Committee members of *24 entreprises et subséquentes*, Focus Group Discussion, Kinshasa, March 2017).

The Interprofessional Guaranteed Minimal Wage Policy of 2008 added a layer of complexity and paradox to the activities of the Labour Department. Mrs Lukiana who implemented this policy was twice minister of labour—from 2001 to 2003 and 2007 to 2008. In 2002, she issued Ministerial Order No. 12/CABMIN/TPS/AR/KF/059/02 on 27 September that pegged the value of the national minimum wage at US$30 per month. The Order, however, was flouted by many employers. In her second term, she persuaded the government to promulgate Ordinance No. 08/040 of 30 April 2008 to fix the minimum wage at US$3 dollars per day but the ordinance too was not enforced by the government (ONEM 2015a).

In September 2008, the Labour Department issued Ministerial Order No. 12/CAB. MIN/ETPS/059/2008, creating a commission charged with preparing the National Programme of Labour and Vocational Training. Although this programme was submitted to the government in 2009, its resolutions were not implemented (Ministère de l'emploi du Travail et de la Prévoyance Sociale 2012).

Two reasons led to the abandonment of these two labour market policies. The first related to several cabinet reshuffles through which new ministers were appointed. These officials embarked on new projects and discontinued the projects initiated by their predecessors. For instance, Labour Minister Mr Nzanga Mobutu, who replaced Mrs Lukiana in 2009, initiated the National Youth-Employment Programme in 2010 while Mrs Lukiana's National Programme of Labour and Vocational Training was pending (Inaka 2020). The second reason was that the cabinet did not provide timely feedback on projects put forward for adoption by the Labour Department(Khondem 2017).

When we suggested these reasons to a manager and four deputy managers in the Labour Department, they spoke passionately about the national labour market policy. The manager, for instance, said,

> We always complain that the government neglects our propositions … It's sad that we don't have a coherent national labour market policy. (Likwalo, Manager, Labour Department, Personal Interview, Kinshasa, February 2017)

In September 2008, the Labour Department tried to tackle the problem of private job placement services that violate labour laws. As some of these placement services operated illegally as outsourcers, Mrs Lukiana issued Ministerial Order No. 006/CAB. MIN/ ETPS/ 062 of 18 September 2008 to clarify the roles of private job placement services (Ministère de l'Emploi du Travail et de la Prévoyance Sociale 2008b). Yet most of the transgressors simply continued outsourcing their operations (Kisimba 2014). Worse still, a report from the National Employment Office indicated that some private job placement services obliged candidates to pay job application fees. There were thus three problems with these agencies: they operated illegally as outsourcers; they took a cut of salaries of the employees they placed; and they unfairly fired their employees (ONEM 2015a). Consequently, on 10 November 2014 Labour Minister Modeste Bahati prohibited 49 of 70 private job placement agencies from operating in Kinshasa (Kisimba 2014). But the prohibition was ephemerally implemented as Mr Bahati was replaced in March 2015. His successor, Labour Minister Willy Makiashi, acknowledged on radio in April 2015 that most of the companies that had been prohibited continued operating (Radio Okapi 2015). Parliament finally passed a law on outsourcing in the private sector on 8 February 2017 to end the practice (République démocratique du Congo 2017).

Our interviewees disclosed a range of additional problems that affected the internal functioning of the Labour Department. For example, politicisation and interference by military and/or political authorities in the department's internal affairs. Routinely, corruption, lack of infrastructure, mismanagement, low wages and staff recruitment based on tribal, ethnic and family affinities affected its functioning. A manager complained that ministers often centralised all the institution's activities in their offices and made decisions without consulting experts. The interference by non-labour actors, he argued, weakened the department in favour of unscrupulous employers:

> There are many cases where decisions of the General Labour Inspectorate in favour of the workers have been cancelled by politicians, the guys of the National Intelligence Agency. (Manager, Labour Department, Personal Interview, Kinshasa, February 2017)

This manager placed the corruption more at the level of the cabinet and said it had been entrenched for many years. He gave the example of an open letter dated 15 June 2015 addressed to the Minister of Labour, Mr Willy Makiasi. In the letter, the author accused a South African private construction company, Redis Construction Afrika, of bribing senior executives of the Labour Department and the National Commission of Employment of Foreign Employees to gain permission to hire more than 1,500 South African and Zambian employees and not a single Congolese (Kabiola 2015; Mambuya 2018).

The Labour Department did not accomplish most of its objectives. Project discontinuity led to overlaps in unachieved policies, poor capacity to enforce laws and the impunity of lawbreakers. The department could not protect the work of the national labour force because of the lack of continuity and partial decision making. The lack of support from the government resulted in unimplemented projects (such as the wage policy). And problems such as corruption, politicisation and interference by non-market actors exacerbated the situation. Effectively, the department's impact on the remaking of the labour market in the DRC actually declined after 2003. If the department itself has had difficulty functioning and contributing to the reconstruction of the job market, it opens the question of where this leaves the institutions under its supervision.

National Council of Labour

The National Council of Labour (CNT) comprises three members including, government, employer organisations and trade unions. According to Articles 224 to 229 of the 2002 Labour Code (République Démocratique du Congo 2002), the duty of the CNT as a consultative organisation is to analyse economic, financial and social issues on the labour market, assess all matters on labour and welfare in the DRC and express its views before adopting laws that impact the labour market. Theoretically, the CNT has always been consulted, as is stated in the preamble of each labour law; that all new bills must pass through the institution. However, our findings suggest that this is not always the case. Some members of the institution are regularly left out in law-making processes. Contradictions among members too such as those on the implementation of the 2017 wage policy resulted in the policy being blocked. At the law-making level, a manager at the National Employment Office and a trade unionist both stated that the promulgation of the 2017 Act No. 17/001 on Outsourcing did not follow the regular procedures. They complained that government did not consult all CNT members and the trade unions in particular (ONEM Manager, Personal Interviews, Kinshasa, June 2017; informal conversation with a trade union Senior Executive, Kinshasa, May 2017). In contrast, a manager of the Congolese Business Federation said this was incorrect and that the 2017 Act followed the regulated procedures:

> I'm telling you that members of the National Council of Labour met and discussed about that law. Our representatives took minutes [at the meetings] ... These unionists are worse than politicians: they were invited but chose not to come! (FEC manager, informal conversation, Kinshasa, June 2017)

An interview with an MP brought even greater clarity on these conflicting opinions on the same issue. For the MP, although Act No. 17/001 theoretically followed the regulated procedures, some controversies on how it was developed needed explaining:

> I saw the minutes of the National Council of Labour meeting without signatures from the members of the many trade unions that I know. ... There was no substantial content: no comments, no recommendations, no suggestions, in short, nothing! It was awkward ... Arrangements had already been made between the cabinet and MPs of the ruling party to adopt it. (Member of Parliament 1, Personal Interview, Kinshasa, June 2017)

These informants provide strong indication of irregularities within the CNT's ' role in labour legislation processes. The engineered absence of the trade unions, clearly a key member of this institution, shows why from a law-making perspective some labour policies might be regarded as ill-conceived. Additionally, the practice of *loi marathon* corroborates the literature (see e.g. Trefon 2011) that depicts the national Parliament as a rubber-stamping machine for the ruling party. This practice also allows us to understand why the law-making processes take longer.

CNT decided to implement the minimum wage policy in November 2017, when all its members agreed on a minimum wage of US$5 per day (Buhake 2017). They proposed that government implement the resolution by 1 January 2018. Government, however, only issued Ordinance No. 18/017 on the Minimum Wage on 22 May 2018. The delay gave members of the Congolese Businesses Federation the opportunity to veto its implementation under the pretext that employers would no longer respect the CNT decision because of the government's delay (FEC 2018).

Overall, internal contradictions among CNT members hindered it from fulfilling its mission. CNT is a hotbed for conflict, despite being a platform for key players in the labour market. This causes delay in the implementation of labour market policies and makes it plausible to argue that the institution was designed to resist the reconstruction of the labour market.

National Institute for Vocational Training

The main responsibility of the National Institute for Vocational Training (Institut National pour la Préparation Professionnelle, INPP) stipulated in Articles 11 to 17 of the 2002 Labour Code is to train jobseekers or new (young or adult) employees in the workplaces (République Démocratique du Congo 2002). It was created through Ordinance Act No. 206 of 29 June 1964. Problems in the way INPP functioned arose in the late 1970s from conditions in the DRC's public administration. In particular, looting in urban areas in 1991 and 1993 and the wars of 1996–1997 and 1998–2002 brought INPP to the verge of collapse. The political context of the transition also did not help INPP alter its course (Inaka 2020). An assistant manager at the institute explained that even in the late 2010s it was not performing well because it had no appropriate equipment, ageing instructors and teaching methods and poor management:

> Demobilised child soldiers needed to be given short-term, job-specific training so that they would have practical skills to find jobs. However, our instructors were under-qualified and unable to respond to the changing needs of the labour market. Many trainees spent their time learning antiquated theories instead of practical techniques. … Our equipment was gone. … Our managers were incompetent. … Employers could not trust us. They stopped sending their employees to attend our training sessions. (Assistant Manager, INPP, Personal Interview, Kinshasa, July 2017)

The situation changed, however, from 2007. At this point, the assistant manager reported, they introduced two measures which improved the INPP's functioning and contributed to its achievements in the labour market. The first measure was to reinforce its relations with external donors and international cooperation agencies such as the Japan International Cooperation Agency (JICA), the Belgian Technical Cooperation and the German Society for International Cooperation. The second measure was to restructure the INPP by upgrading the skills of instructors and the standard of the training programmes, renovating its buildings and constructing new facilities (Assistant Manager, INPP, Personal Interview, Kinshasa, July 2017). As a result of the renewal and expansion of the DRC's international relations, JICA sent INPP instructors for retraining in Dakar in 2008. On their return, they trained their colleagues in DRC (JICA 2011). The donors also funded the renovation and construction of facilities. INPP could, therefore, upgrade its training programmes and improve performance (Inaka 2020).

Most informants in this study had positive views of the INPP's impact on the labour market and disclosed that it had even led to some surprisingly positive results. For example, many employers in construction and telecommunication prefered hiring INPP-trained technicians to engineering graduates of DRC colleges or universities. A manager from the Congolese Business Federation, for example, stated that employers found INPP-trained technicians more competent:

> If we have to say it, there are some differences in practical skills between an engineer from ISTA (*Institut supérieur de techniques appliquées*, Higher Institutes of Applied Technologies) and someone from INPP. In the labour market, employers prefer someone from INPP. They have demonstrated their competencies. (Manager, FEC, Personal Interview, Kinshasa, December 2016)

An Assistant Manager at the INPP added:

> As our ex-learners prove knowledge, many companies invite us to train their employees. Even the FEC recognises the skills of our learners today. I want to show you an article in which the CEO of the FEC congratulates us. Look what is written here: 'If [in the past] the INPP was unable to meet the expectations of companies', said the CEO, 'today the institute is able to do it'. (Assistant Manager, INPP, Personal Interview, Kinshasa, July 2017)

However, the 'good management' of the INPP has not spared it from politicisation and the ignorance of non-market actors. The pressures of politicians, military and/or security agents are strongly linked with corruption. The Assistant Manager was very candid when describing the pressure that *Bakonzi ya Congo* [Congo's highest authorities] applied on INPP Director General (DG):

> They put pressure on our DG to give them money ... Anyway, he must face the music because he was informed that he had to fund *bakonzi ya Congo* ... It's their deal. As he was the one who negotiated that matter, he is the one who has to deal with it. That's what happens in the Congo. (Assistant Manager, INPP, Personal Interview, Kinshasa, July 2017)

By increasing it functional capacity, INPP survived collapse during the transition period as acknowledged by employers, employees, international partners and the media. Collaboration between international partners and local actors contributed to institutional strengthening and capacity building in this respect. This positive outcome challenges the literature that suggests that vocational training programmes fail in post-war countries. Although such programmes failed in the DRC during the transition period, they turned out successful from 2008.

National Employment Office

The National Employment Office (Office National de l'Emploi, ONEM) remained inactive in the first decade of its existence (2002–2012) because of the political context of the transition period and the Muzito government. ONEM constitutes a governing board, a directorate general, a board of auditors and a directorate board. Its key mission is to connect job seekers and employers (République Démocratique du Congo 2002). A couple of ONEM's achievements are noteworthy for the period 2012 onwards namely, introduction of the *Programme Emploi Diplômé* (PED) (job programme for graduates) in 2015 and the recognition of the right to approve labour contracts (ONEM 2015b). ONEM helps unemployed graduates gain work experience through internships and has placed 1,000 graduates in internships per year since 2015. Each trainee receives US$100 per month, paid through a grant from the African Development Bank (Radio Okapi 2015). An assistant manager at ONEM considered the PED programme its major achievement:

> At least 30% of the trainees end up being hired by these companies. … This is something we appreciate. Because, you see, honestly, before the PED, it was impossible for us to assist even 2% of job candidates per year. (Assistant Manager 1 and Assistant Manager 2, ONEM, Focus Group Discussion, Kinshasa, March 2017)

A manager criticised the PED programme, however, for focusing on numbers of trainees rather than the improvement of their skills. For him, the trainees are placed in useless job positions. Critically, he claimed that the recruitment of these trainees was often based on tribalism and corruption (Manager, ONEM, Personal Interview, Kinshasa, June 2017).

ONEM also helps employees by assessing and approving labour contracts. An assistant manager explained it this way:

> When we started, some employers used to send labour contracts written only in English, though they knew that our official language is French. We refused to approve them … Now they send both an English and a French version. (Assistant managers, ONEM, Focus Group Discussion, Kinshasa, March 2017)

Despite these achievements, a manager explained, ONEM was still only fulfilling (and this partially) two out of ten of its tasks, namely assessing and receiving labour contracts and informing and/or training jobseekers. To him, ONEM could not satisfy both employers and job seekers, and job seekers therefore no longer consulted it. Employers, on the other hand, were dissatisfied with most job candidates ONEM supplied (Manager, ONEM, Personal Interview, Kinshasa, June 2017). Similarly, two employees expressed disappointment with ONEM because it did not help them find jobs. Mr Lubudi explained:

> I registered at ONEM in 2013. I was impressed by their advice … They told me they will call. They have never done it. I followed up, but it was always the same song: 'We will call you' … I don't trust them. (Mr Lubudi, Supervisor, New Sarah, Personal Interview, Kinshasa, April 2017)

A marketer employed at the same company as Mr Lubudi added:

> They promised a job. Since then, the job was never given … ONEM does nothing. Outsourcers are even better than that ONEM. (Marketer, New Sarah, Personal Interview, Kinshasa, April 2017)

Another problem arises from the bureaucratisation of hiring processes. Legally, employers must inform ONEM of any vacancies before advertising them. ONEM then has 30 days to find the right candidates for the positions. Only if it fails do employers have the right to recruit candidates directly, as elaborated in Box 5.1.

Box 5.1 Hiring Procedure for Nationals in the DRC
Legal framework:

- Law No. 015/2002 of 16 October 2002 on the Labour Code (Articles 203–207);
- Enactment No. 74/098 of 6 June 1975 on the Protection of the National Workforce.

Procedure:

- Declaration of vacancies at ONEM;
- Hiring of a worker shall be subject to a contract in writing and concluded for a fixed or indefinite period. This contract specifies explicitly the nature and class of employment, the rate of pay, the benefits to which a worker is entitled, and other mandatory clauses imposed by the Labour Code;
- Presentation by the applicant of all documents required of him by the company, establishing a track record prior to hiring;

After hiring a worker, the employer must declare it to the office of the Ministry of Labour.

Source National Agency for the Promotion for Investment (2016)

The legal procedure outlined in Box 5.1 is very laborious, as ONEM databases do not often contain anyone with the skills that employers require. Consequently, employers often do not follow procedure, preferring to pay fines if caught (Manager, ONEM, Personal Interview, Kinshasa, June 2017). ONEM is also quite ineffective in issuing information on employment opportunities. It, therefore, fails to achieve its main objective of eradicating irregular recruitment, as a manager there argued (Manager, ONEM, Personal Interview, Kinshasa, April 2017). This claim is consistent with the finding of the National Institute of Statistics that 79.9% of all jobs are obtained via personal networks (through family and political relations), 24.4% through wider networks (through friends and acquaintances), and only 0.7% through ONEM (INS 2014).

To sum up, ONEM is weakened by politicisation, corruption, mismanagement and the interference of non-market actors in its internal affairs. According to an assistant manager, the interference by several political, military and secret service agents in the affairs of ONEM constitutes an incurable administrative pathology (Assistant Manager, ONEM, Personal Interviews, Kinshasa, March 2017). Corruption is a major drain, with several assistant managers stating that the organisation was filled with thieves, from top to bottom (Assistant manager, ONEM, Personal Interview, Kinshasa, March 2017). The media has indeed, for example, reported that ONEM's Director General, Mrs Angélique Kikudi, has embezzled funds provided by the African Development Bank meant to be stipends for trainees in a Jobs for Graduates Programme (Kazadi 2016). It would thus be difficult to suggest that ONEM has contributed positively to the reconstruction of the labour market in the DRC since 2003 as the highlighted constraints prevent it from fulfilling its mandate.

National Social Security Fund

The mission of the National Social Security Fund (Caisse Nationale de Sécurité Sociale, CNSS), according to Decree No. 18/027 of 14 July 2018 on the Creation, Organisation and Function of the National Social Security Fund, includes collecting social security contributions for employees from their employers, paying social benefits to retired workers and to workers disabled by work-related accidents, and paying benefits to families, including family allowances and maternity allowances (République Démocratique du Congo 2018).

A tripartite governing board composed of representatives of the government, employers, and employees; a directorate general; a board of auditors; and a board of directors runs CNSS (Caisse Nationale de Sécurité Sociale 2019). The organisation was first established as the National Social Security Institute through Decree-Organisational Act of 29 June 1961 (République démocratique du Congo 2003). It was later transformed into the CNSS following the enactment of Law No. 16/009 of 15 July 2016, which set the rules for the general social security scheme (Caisse Nationale de Sécurité Sociale 2019).

In July 2011, the CNSS opened up membership of its welfare schemes to informal sector workers (Radio Okapi 2011). It also extended membership to Congolese who worked in diplomatic missions in the country, in NGOs, and international organisations as contract workers, self-employed, senior executives parastatal companies, or shareholders in private companies (Mbuyi 2016). In 2017, it standardised access to retirement at the age of 65 years for both men and women (Le Phare 2018). Still, CNSS had challenges convincing employees and employers to subscribe to its welfare schemes. A manager in the organisation explained that informal workers worked in precarious conditions because of the lack of social security coverage in their sector and this made it necessary for them to have access to welfare schemes. The CNSS project failed, however, because informal workers as well as workers in NGOs, international organisations, contract workers and self-employees refused to subscribe to its schemes (Manager, CNSS, Personal Interview, Kinshasa, June 2017).

Besides, two major problems had prevented them from paying pensions regularly during that period. First, the CNSS uses the contributions of current members to pay retirees. But because many current members did not pay their contributions because of uncertain political circumstances and wars, the CNSS did not have the cash flow to pay pensions. Second, the CNSS uses rent from a large estate it owns to pay out pensions. But many tenants had not paid their rentals. Additionally, many of the properties were hijacked by political and military authorities who even illegally sold some of the buildings (Assistant Manager, CNSS, Personal Interview, Kinshasa, June 2017).

This explanation was only partially true, however. In speaking of how CNSS almost failed to comply with the ILO's recommendation on increasing funds allocated to welfare expenditure, the Assistant Manager revealed that,

> Since 2011, the CNSS allocates 60% of its revenues to pay for pensions and 40% for running its administration, due to the pressure from the ILO. Before 2011, 20% was dedicated to pay for pensions. Regardless of the increase, the ILO wants us to comply with international norms. I mean, to put aside 80% to pay for pensions … it's difficult because the state doesn't fund us. (Assistant Manager, CNSS, Personal Interview, Kinshasa, June 2017)

With these words the informant unwittingly pointed to the real source of the tension between the CNSS and retirees. This was that the institution used to take the lion's share of its revenues (80%), while allocating crumbs to pay out pensions for retirees (20%). There is also reason to believe that the CNSS did not appreciate the pressure exerted by the ILO because it wanted to retain the status quo. That is why they decided to allocate only 60% to pensions instead of respecting international standards pegged at 80% of income.

The CNSS manager also pointed to corruption affecting the running of the organisation. He particularly blamed this on the politicisation of the institution. He gave no concrete details or examples, however. Some information exists in the public domain, however, that the organisation's Deputy Director General and financial manager stole US$163, 000 in May 2013 from the organisation (Radio Okapi 2013).

Nevertheless, the data shows that the CNSS did act to fulfil its mandate: to collect social contributions from employees or via their employers, and to provide welfare benefits to its beneficiaries. Most of its actions were oriented towards collecting funds, and rather few towards providing welfare benefits. The tendency to increase its pool of income is demonstrated by its desire to expand the categories of employees in the membership of its social security schemes. However, it has implemented fewer concrete measures to pay out welfare benefits. Actually, it seems as if the CNSS wants one thing and the opposite—while it fights to increase the number of members, it displays the inability to deliver services to current members. Overall, pensioners continue to suffer and based on the foregoing, the impact of the CNSS on the reconstruction of the labour market in DRC has been a negative.

National Agency for the Promotion of Investments

The National Agency for the Promotion of Investments (Agence Nationale pour la Promotion des Investissements, ANAPI) remained dormant for eight years before it became fully operational in 2009. Its specific mission, according to Decree No. 09/33 of 8 August 2009, which regulates the status, organisation and functioning of the organisation includes promoting a positive image of the DRC, advancing specific investment opportunities, improving the business climate in the country and providing support to investors who establish or expand economic activities in the country (République démocratique du Congo 2009).

ANAPI is managed by a governing board, a directorate general and an accreditation council that seek to accomplish these aims (ANAPI 2016). Its best-known achievement in the demand for labour was the creation of a 'one-stop shop' with associated improvements in business legislation. The creation of the one-stop shop in 2012 resulted in the simplification of processes required to set up a business in the DRC (ANAPI 2018). It also reduced the bureaucratic requirements for new investments when it unified many public agencies and services (van Impe 2012). In 2013, it contributed towards legal reforms that reduced the cost of investment processes, and in 2014 introduced measures to assist investors in setting up new businesses (ANAPI 2018). Mr Gbadolite, an ANAPI manger, explained:

> Before the creation of one-stop shop, investors had to look for documents at many institutions like the INSS, the INPP, the Department of Justice, etc. They bribed to accelerate processes or reduce the costs. But today that's no longer necessary. It is easier to get everything at the same place. (Gbadolite, Manager, ANAPI, Personal Interview, Kinshasa, March 2017)

A second ANAPI manager, Mr Makanza, added:

> To start a business took three months in the past. Now we have shortened that time from three months to three days. (Makanza, Manager, ANAPI, Personal Interview, Kinshasa, March 2017)

Concerning the level of improvement of the business climate, Mr Gbadolite stated:

> We are trying to improve the business climate, and our ranking on the Doing Business reports … The problem is that the more we improve, the more other countries are also improving. So we are still ranked among the lowest. (Gbadolite, Manager, ANAPI, Personal Interview, Kinshasa, March 2017)

Regarding labour incorporation processes, Mr Gbadolite argued that ANAPI paved the way for ONEM:

> We attract investors who create job opportunities. Then it is the role of ONEM to supply employees to these investors. (Gbadolite, Manager, ANAPI, Personal Interview, Kinshasa, March 2017)

But ANAPI has also found itself struggling with delays in decision making in other branches of government, corruption, encroachment of non-market actors on its internal affairs, and politicisation. As one manager complained:

> Often the government can change the law, but putting it into practice does not follow. And that affects investors, when they come: they know that the law says this or that, but on the ground, it is the opposite that happens. (Makanza, Manager, ANAPI, Personal Interview, Kinshasa, March 2017)

Mr Gbadolite expanded on this:

> Insecurity in some parts of the country, such as in the East, is a real problem. There are also all the ills that plague the country, such as corruption, the politicisation of all the institutions, and the lack of political good will and good governance. (Gbadolite, Manager, ANAPI, Personal Interview, Kinshasa, March 2017)

The impact of ANAPI on the labour market is open for debate. First, the facilitation of investment processes and the improvement of the business climate might be regarded as progress in the sphere of the demand for labour. As investments and new businesses increase, job opportunities might increase too in the labour market. ANAPI's statistics show an increase in the number of companies in the DRC between 2013 and 2017 from 1809 to 6039 (ANAPI 2018). Second, attracting investors does not necessarily lead to a reduction in unemployment. A report by the DRC's National Institute of Statistics shows that the rate of unemployment remains over 80% and that access to formal employment is restricted in the country (INS 2018). These statistics are debated because ANAPI only provides information on the number of companies created but does not explain in which sector or industry these companies are located. It is thus unclear whether these companies are one-man shows, small-to-middle-scale employers, outsourcers, or consulting companies, making their labour absorption capacities unclear. Nor do the statistics show whether these companies hire Congolese or foreign employees. It is, therefore, worth considering that ANAPI's efforts have less visible impacts on labour absorption. It can be inferred that a huge gap exists between increasing the number of investors and creating job opportunities in a post-war labour market. Overall, ANAPI has been trying to improve business and the demand for labour. Yet it faces issues in implementing adequate labour demand policies.

Labour Market Institutions in the DRC in Comparative Perspective

Overall, labour market institutions design and implement labour projects. Yet, as shown above, few of these projects have achieved the expected results. The CNT's 'national programmes of employment and wages were delayed for years. This is also true of its minimal wage policy. ONEM was slow to reach its objectives of linking labour supply and demand. And at the CNSS, several initiatives remain less than effective from the perspective of retired people. On the other hand, the INPP and ANAPI have made some progress. But they are two tiny entities in a dysfunctional institutional system. As such, it is difficult for them to celebrate the status of being an exception to the general rule.

These findings can be explained with the help of Peck's (1996: 72) notion of the regulatory dilemma of labour market institutions. For him, the labour market is full of contradictions which cannot easily be resolved by institutions. Though there are potential institutional responses, they rarely provide absolute solutions to regulatory dilemmas. It can thus be argued that labour market institutions in the DRC struggled to overcome regulatory dilemmas. They were slow to address labour market problems efficiently due to several interconnected, contradictory and conflicting political, social and economic factors. Indeed, all these facts restricted the influence that labour market institutions could have had on the post-war reconstruction of the country's labour market.

Conclusion

This chapter examined the role of public labour market institutions in the remaking of the labour market in the DRC to enhance the understanding of the roles that such institutions play in post-war labour market recoveries. It has shown that public institutions like the Labour Department of the Government of the DRC, the National Employment Office, National Social Security Fund and the National Council of Labour are weak and labour market policies in these institutions are poorly implemented or not implemented at all and ineffective. The institutions face political, socioeconomic and cultural challenges. Political factors negatively affect the institutions, and this was particularly evident between 2003 and 2006 during the transition. Other challenges facing the institutions include corruption, public maladministration and mismanagement. In line with Peck (1996)'s notion of labour market regulatory dilemmas, labour market institutions in the DRC struggled to overcome the regulatory dilemmas they faced. As a result, they did not effectively and satisfactorily contribute to the remaking of the labour market.

References

ANAPI, Agence Nationale pour la Promotion des Investissements. 2016. "Organisation." *Accueil, a Propos, Organisation*, 3 February 2016. Kinshasa ANAPI. Retrieved 18 April 2019 (https://www.investindrc.cd/fr/anapi/organisation).

ANAPI, Agence Nationale pour la Promotion des Investissements. 2018. "Facilite D'entreprendre Les Affaires En Rdc." Kinshasa ANAPI. Retrieved 1 June 1 2019 (https://investindrc.cd/fr/guide-de-l-investisseur/facilite-d-entreprendre-les-affaires-en-rdc).

Buhake, Dina 2017. "Rdc: Le Smig Passe De 1.680 a 7.075 Fc/Jour, Soit 5 Usd/Jour." *Forum des As*. Kinshasa.

Caisse Nationale de Sécurité Sociale, la CNSS. 2019. "A Propos De La Cnss." Kinshasa Caisse Nationale de Sécurité Sociale (http://www.cnss.cd/a-propos/CNSS).

FEC, Fédération des entreprises congolaises. 2018. "Préoccupations Des Membres De La Fec Suite Au Décret Portant Sur Le Smig." Retrieved 15 June 2018 (http://www.fec-rdc.com/index.php/actualites/24-actualite1/212-preoccupations-des-membres-de-la-fec-suite-au-decret-portant-sur-le-smig).

Inaka, Saint José. 2020. "Post-War Labour Market Reconstruction: The Case of the Democratic Republic of the Congo." Doctoral dissertation, Sociology, University of Pretoria Pretoria.

INS, Institut National de la Statistique. 2014. "Rapport Global, Enquete 1–2–3. Resultats De L'enquete Sur L'emploi, Le Secteur Informel Et Sur La Consommation Des Menages." Vol. *Rapport Global*. Kinshasa: INS.

INS, Institut National de la Statistique. 2018. *Enquete Avec Questionnaire Unifie a L'indicateur De Base De Bien Etre, Rapport De L'enquete*. Kinshasa: World Bank Group.

JICA, Japan International Cooperation Agency. 2011. "Getting the Congo Moving Again." Retrieved 25 December, 2017 (https://www.jica.go.jp/english/news/focus_on/senegal2011/senegal_01.html).

Kabiola, Albert Bikilwa Nfika 2015. "Lettre Ouverte À Monsieur Willy Makiashi, Ministre De L'emploi, Travail Et Prévoyance Sociale." Actualite Congolaise et Africaines. le Congolais.cd. Retrieved 14 April 2019 (https://www.lecongolais.cd/lettre-ouverte-a-monsieur-willy-makiashi-ministre-de-lemploi-travail-et-prevoyance-sociale/).

Kazadi, Caleb. 2016. "Programme Emploi Diplômé (Ped): Qui Bloque La Bourse De La Bad ?" *Actualite.cd*. Kinshasa Groupe Next Corp (https://actualite.cd/2016/12/15/programme-emploi-diplome-ped-qui-bloque-la-bourse-de-la-bad).

Khondem, Eugène. 2017. "Congo-Kinshasa: Sans Croissance Économique Inclusive Locale—La Fête Du Travail Commémorée Dans Une Forte Controverse." Pp. 1–2 in *La Prospérité*. Kinshasa: La Prospérité.

Kisimba, Dorian. 2014. "Modeste Bahati Ferme 49 Entreprises Privees De Placement Non En Regle." *7SUR7.CD*. Kinshasa 7sur7.CD.

Le Phare. 2018. "Rdc: Inss Devient Caisse Nationale De Sécurité Sociale (Cnss)." *Le Phare*. Kinshasa: le Phare.

Mambuya, Adelard Obul'Okwess. 2018. "En Rdc, Les Inspecteurs Du Travail N'ont Que Leur Volonte." *Actualite politique* Actualite. cd (https://actualite.cd/index.php/2018/03/10/en-rdc-les-inspecteurs-du-travail-nont-que-leur-volonte-brand-content).

Mbuyi, Dieudonné. 2016. "Organisation Du Régime Général De Sécurité Sociale En Rdc: Agnès Mwad Et Les Travailleurs De L'inss Saluent La Loi Promulguée." Retrieved 7 Juilley 2018 (https://7sur7.cd/organisation-du-regime-general-de-securite-sociale-en-rdc-agnes-mwad-et-les-travailleurs-de-linss-saluent-la-loi-promulguee/).

Ministère de l'Emploi du Travail et de la Prévoyance Sociale. 2005a. *Arrêté Ministériel N° 12/Cab.Min/Tps/117/2005a Du 26 Octobre 2005 Fixant La Durée Et Les Conditions De Préavis*. Kinshasa Journal officiel.

Ministère de l'Emploi du Travail et de la Prévoyance Sociale. 2005b. *Arrêté Ministériel N° 12/Cab.Min/Tps/116/2005b Du 26 Octobre 2005 Fixant Les Modalités De Licenciement Des Travailleurs*. Kinshasa Journal officiel.

Ministère de l'Emploi du Travail et de la Prévoyance Sociale. 2008a. *Arrete Ministeriel No. 12/Cab. Min/Etps/059/2008a Du 18 Septembre 2008 Portant Creation, Organisation Et Fonctionnement De La Commission Chargee De La Formulation Du Programme National Pour L'elaboration Et La Mise En Oeuvre De La Politique De L'emploi Et De La Formation Professionnelle*. Kinshasa Journal Officiel.

Ministère de l'Emploi du Travail et de la Prévoyance Sociale. 2008b. *Arrêté Ministériel N° 012/Cab.Min/Etps/062/08 Du 18 Septembre 2008 Fixant Les Conditions D'ouverture, D'agrément Et De Fonctionnement Des Services Privés De Placement*. Kinshasa: Journal Officiel.

Ministère de l'emploi du Travail et de la Prévoyance Sociale. 2012. *Programme Pays Pour Un Travail Décent (Pptd) 2013–2016 Du Bureau De Pays De L'oit Pour L'afrique Centrale*. Congress.

National Agency for the Promotion for Investment. 2016. "Hiring and Dismissal of Workers". ANAPI. Retrieved 27 April 2019 (https://www.investindrc.cd/en/procedures/other/152-hiring-and-dismissal-of-workers).

ONEM, Office National de I'Emploi. 2015a. *Études Des Tendances Du Marché De L'emploi En République Démocratique Du Congo. Réponse Au Vice Premier Ministre, Ministre De L'emploi, Du Travail Et De La Prévoyance Sociale*. Kinshasa: ONEM.

ONEM, Office National de I'Emploi. 2015b. *Rapport National Sur L'emploi. Base Sur Le Modele Adopte Par La* SadcCongress.

Radio Okapi. 2011. "Kinshasa: L'inss Vulgarise Ses Prestations Sociales." INSS sécurité sociale En bref Kinshasa National. Retrieved 26 Juin 2017 (https://www.radiookapi.net/regions/kinshasa/2011/07/05/kinshasa-l%25e2%2580%2599inss-vulgarise-ses-prestations-sociales-2).

Radio Okapi. 2013. "Rdc: Le Directeur Général Adjoint De L'inss Accusé Du Détournement De 163 000 Usd." Retrieved 18 June 2019 (https://www.radiookapi.net/societe/2013/05/08/rdc-le-directeur-general-adjoint-de-linss-accuse-du-detournement-de-163-000-usd).

Radio Okapi. 2015. "Willy Makiashi: «Une Confusion Totale Délibérée S'est Installée Dans Les Attributions Des Maisons De Placement»." emploi Placement RDC Sous-traitance Willy Makiashi Émissions Interview. Retrieved 9 June 2018 (https://www.radiookapi.net/emissions-2/linvite-du-jour/2015/04/01/willy-makiashi-une-confusion-totale-deliberee-sest-installee-dans-les-attributions-des-maisons-de-placement).

République Démocratique du Congo. 2002. *Loi N°015/2002 Portant Code De Travail*. Kinshasa: Journal Officiel de la Republique Democratic du Congo.

République démocratique du Congo. 2003. *Les Codes Larcier: RéPublique DéMocratique Du Congo*. Bruxelles et Kinshasa: Larcier; Afrique éditions.

République démocratique du Congo. 2009. *Décret No. 09/33 Du 08 Août 2009 Portant Statuts, Organisation Et Fonctionnement De L'agence Pour La Promotion Des Investissements, En Sigle «Anapi»*. Kinshasa: Jounral Officiel de la République démocratique du Congo.

République démocratique du Congo. 2017. *Loi No. 17/001 Du 08 Fevrier 2017 Fixant Les Regles Applicables a La Sous-Traitance Dans Le Secteur Prive*. Kinshasa: Journal Officel.

Trefon, Theodore. 2011. *Congo Masquerade: The Political Culture of Aid Inefficiency and Reform Failure*. London: Zed Books Ltd.

van Impe, Andreas. 2012. *Investing in the Democratic Republic of the Congo. A Legal Guide*. Brussels: De Wolfs and Parterns.

CHAPTER 6

Post-war Labour Market Reconstruction, Private Labour Market Institutions and Non-market Actors

This chapter builds on the foundation laid in the first part of this book and continues the discussion of the impacts of private labour market institutions and non-market actors on post-war labour market reform and recovery from Chapter 5. The focus remains on the Democratic Republic of the Congo (DRC) and the analysis draws on Peck's notions of the labour market as an arena of contestation, conflict, contradiction and complexity. With this and data gathered from personal interviews and focus group discussions, the chapter suggests that a correlation exists between the internal weaknesses of public labour market institutions in the DRC and those imposed by external factors. Further, the private sector rather weekly contributes to the remaking of the labour market. This is because of conflicts of interest between actors and the quasi-domination of employer organisations. The chapter first examines the contribution of the private sector labour market to the reconstruction of the labour market since independence and then discusses the involvement of non-market actors in those institutions and how the actors weaken the institutions as well as weaken the contribution of the institutions to post-war labour market reconstruction.

S. J. Inaka et al., *The Reconstruction of Post-War Labour Markets in The Southern African Development Community*,
https://doi.org/10.1007/978-3-031-37357-2_6

Private Labour Market Institutions

The Congolese Business Federation and trade unions have significantly contributed to the labour market of the DRC and the post-war reconstruction of the market. The influence of these institutions on the labour market extends from the colonial period, President Mobutu and Laurent-Désiré Kabila's respective regimes, up to the post-war period in the 2020s.

Brief Historical Overview of Employer Organisations

Besides holding companies or trusts and chambers of commerce that operated in the DRC and had notable influence on its labour market from the colonial period up to independence, interconnected employer organisations had noteworthy influence on the labour market too. Examples of influencial chambers of commerce include the Chamber of Leopoldville, Chamber of Katanga and Chamber of Stanleyville while examples of holding companies include, *Société Générale*,[1] *Compagnie pour le Commerce et l'Industrie du Congo*,[2] *Groupe Empain*,[3] and Lever Brothers[4] (Buelens and Marysse 2009, Piret 2011). Each chamber of commerce operated under the supervision of the National Federation of Chambers of Commerce, Industry, and Agriculture (Matenda 2002).

The National Federation of Chambers of Commerce formed part of the 'colonial trinity', 'bloc colonial' or the 'tripartite alliance', which

[1] The *Société Générale* was the largest Belgian holding company and main shareholder of *l'Union Minière du Haut Katanga* (copper and other metals), *la Société Internationale Forestière et Minière du Congo* (diamonds and timber), *la Compagnie du Chemin de fer du Bas-Congo au Katanga* (railways and construction) and *la Banque d'Outremer* (banking) (Buelens and Marysse 2009).

[2] The *Compagnie pour le Commerce et l'Industrie du Congo* was a holding company with several daughter companies, such as the *Compagnie du Chemin de Fer du Congo* (railways), the *Compagnie des Magasins Généraux* (retail), the *Société Anonyme Belge pour le Commerce du Haut Congo* (import–export), the *Compagnie des Produits du Congo* (import–export), and the *Companie du Katanga* (mining) (Buelens and Marysse 2009).

[3] The *Groupe Empain* owned the *Compagnie des Chemins de Fer du Congo Supérieur aux Grands Lacs Africains* (railways) and was the main shareholder of the Banque du Congo Belge (banking) (Buelens and Marysse 2009).

[4] Lever Brothers owned the *Huileries du Congo Belge* (palm oil). As of March 2023, Unilever was a joint venture between Unilever PLC (UK) and Unilever NV (The Netherlands), making it a British-Dutch multinational corporation.

comprised the state, Roman Catholic Church missions and big business in the Belgian Congo (Nzongola-Ntalaja 2002). The alliance made the labour market interdependent and mutually supportive (Reybrouck 2014). The Roman Catholic Church missions supplied the state and private companies with workers (Yates 1978). In return, the state and private companies subsidised the church. The state ensured that the labour force was available for private companies, whereas the latter paid taxes and contributed to construction projects run by the government (Young and Turner 1985). Employer organisations had the assurance of protection of their interests by the colonial administration.

Given this close relationship, employer organisations were successful at lobbying the state to enforce labour market policies aimed at keeping labour cheap (Yelengi 2000). This dominant position was only challenged in the late 1950s with the struggle against colonialism, both locally and internationally, against the background of the developing bipolar international system. After WWII and the signing of the Charter of the United Nations in 1945, the UN's support for the right to self-government or self-determination had considerable impact on Belgian colonial politics. Although Belgium at first insisted on maintaining its control over the Congo (Young and Turner 1985), it finally yielded to pressure by the superpowers, especially the USA, which demanded the colony's independence (Nzongola-Ntalaja 2002).

At the same time, the formidable rise of African liberation movements against colonialism also influenced the DRC (Ndaywel 2006). Emerging political elites and trade unionists in the DRC organised protests, strikes and boycotts against the Belgian administration and against private employers (Nzongola-Ntalaja 2002). In Belgium, the entry of the Belgian Socialist Party and the Liberal Party into government in 1945 curtailed the supremacy of the colonial bloc and, especially, its employer organisations (Ndaywel 2006). Up to then, almost all the Belgian colonial ministers who had managed the Belgian Congo and had supported the trinity of state, church and private companies, were Catholics (Vanthemsche 2012). The rise into power of socialists and liberals, who were, respectively, anti-industrial and anticlerical, diminished the privileges of the Catholic Church and private companies in the Congo (Ndaywel 2006).

Fearful of the coming political order, employer organisations devised schemes to retain control of the Congolese economy after independence.

They created joint stock companies, transformed their productive businesses into financial investments, and moved their companies out of the DRC's jurisdiction into Belgium's. Thus, all the mining, agricultural, transport, manufacturing, construction and trade sectors remained under the control of these organisations (Depelchin 1992).

During the early phase of Congo's independence in the 1960s, local and foreign employer organisations competed with each other and against the state. Emerging Congolese employers urged the new political authorities to bestow privileges upon their businesses. But the political authorities could not act against former employer organisations because they were now under Belgian jurisdiction (Depelchin 1992). Consequently, the economic power relationships that had existed in the labour market during the colonial period extended.

The situation started to change in 1972, when the Mobutu regime created the *Association Nationale des Entreprises du Zaïre* (National Association of Zairian Enterprises, ANEZA) through the Ordinance Law No. 72/028 of 14 September 1972 to control all the country's employer organisations and developed the *Zaïrianisation* policy to expropriate foreign-owned businesses in 1973 (Matenda 2002; Ndaywel 2006). ANEZA's main stated mission was to protect employers' interests (as an organisation of employers), and to promote private sector activities in the Congo (as a chamber of commerce) (Ndaywel 1998). The rationale, for the Mobutu regime, was to fully control the national economy and thwart the emergence of a national bourgeoisie. The reality, however, turned out very different, as the new Congolese owners of companies expropriated from foreign owners bankrupted them in due course (Willame 1995).

After the Mobutu regime collapsed in 1997, ANEZA was renamed *Fédération des Entreprises du Congo* (Congolese Business Federation, FEC) (Matenda 2002) but the wars of 1996–1997 and 1998–2002 prevented it from attracting investors or creating jobs (Bossekota and Sabiti 2001). The FEC only began to implement labour market reconstruction measures after 2003, in the post-war era.

Generally, employer organisations have tried to use the government to protect their business interests and exploit workers starting from the colonial period to the 2020s. Though the identities of employer organisations have changed since the Mobutu era, their strategies survived into the post-war period.

The Congolese Business Federation in the Post-war Period (2003–2018)

The FEC is a chamber of commerce and a union for its members. Its main mission is to promote and defend the interests of its member companies (FEC 2011). It comprises a general assembly, a governing board, a management committee, an auditors' committee and 15 specific chambers (FEC 2015).

The FEC's work starting from the early 2000s[5] included *inter alia* drafting new labour laws and regulations; addressing matters concerning the employment of foreign workers in the DRC's mining industry; creation of a centre for training members in Congolese business law, fiscal law, labour market legislation and economic matters; and the protection of employers against over-taxation and fiscal harassment under the Pollution Law. The FEC faced significant political, administrative and economic challenges in its work that impeded its functioning.

As the main employer organisation in the DRC and a key member of the Congolese National Council of Labour, the FEC helped draw up and issue the 2002 Investment Code and urged the government to bring ANAPI into existence. Mr. Robert, an FEC manager, explained how his institution influenced that process:

> We urged the state to review and accelerate investment reforms so that companies abroad can be attracted, as they are to other countries ... All these actions led to the enactment of the 2002 new Investment Code, and the creation of the ANAPI. (Manager, FEC, Personal Interview, Kinshasa, December 2016)

The changes Mr. Robert talks about only became operational towards the end of the 2000s, because of the rules governing power sharing during the transition government. The fact that ANAPI became operational during the Muzito period (2008–2011) and under President Kabila's control indicates the FEC's capacity to pressure the government to act in its favour (Trefon 2010).

The FEC also participated in the *Éducation au Katanga* (Education in Katanga, EDUKAT) programme to help resolve conflicts over the employment of foreign workers in Katanga. EDUKAT is a set of projects aimed at matching up vocational training curricula for young Congolese

[5] For insight into the FEC's tasks and projects not related to the labour market, see its webpage (http://www.fec-rdc.com/index.php/notre-federation/nos-realisations).

with labour market needs initiated by the Belgian Technical Cooperation (CTB) in Katanga province in 2013. A manager at the FEC explained the reasons underlying this decision:

> It is the FEC that hires people from universities or colleges. … It was in that context that the FEC was invited by the CTB to participate in EDUKAT. … It was necessary to analyse the links between recruitment and the training programmes. The fact is that today the sectors of industrial technology and mining lack the required engineers … Modern technologies [are] used today but the DRC's educational programme does not allow us to train these engineers. (Manager, FEC, Personal Interview, Kinshasa, December 2016)

A Legal Advisor at the FEC added that the EDUKAT programme resolved labour law conflicts and xenophobic tensions over labour:

> At that time, companies resorted to international expertise even for jobs that our laws say that foreigners should not perform. But companies were forced to look for foreign expertise. For example, you see the new big machines, you need specialists to drive them. There were no Congolese able to use them. Those hi-tech machines used nowadays in mining could not be used by Congolese. … Those politicians, trade unionists, journalists manipulated the locals, traditional chiefs of Katanga. … The Malaysian, Indonesian engineers were afraid to go to work. … Xenophobic tension was rising. … Thus, the CBT solicited the FEC to assist in the programme. (Legal Advisor, FEC, Personal Interview, Kinshasa, March 2017)

This response seems to suggest that employer organisations can help resolve what Segatti (2015) considers mobilisation based on identity or autochthony[6] in labour. Katanga Province was apparently a stronghold of such mobilisations from the early 1960s and during the Mobutu era (Inaka 2014). In 2014, Mr. Kahozi, a former provincial minister of education, confirmed that the EDUKAT programme solved several employment-related problems that had led to ethnic or xenophobic tensions in Katanga (Kahozi 2014).

[6] Autochthony is the right to make a claim of land ownership and access to state services and benefits (Segatti 2015: 15).

The FEC also established a training centre for employers and staff in 2014 (FEC 2019). Its manager explained how the centre improved employer familiarity with legal, fiscal, labour and economic matters:

> The Congolese Business Federation trains its members — meaning employers and their staff — in the fields of taxation, economy, and business management of the Congo. We have a training programme on calculations of redundancy payments, accounting, etc. So, these are key areas that interest companies. (Manager, FEC, Personal Interview, Kinshasa, December 2016)

The FEC manager pointed to the fact employers gained insights from FEC training that helped them deal with issues like tax and a poorly functioning bureaucracy (FEC 2019). However, a report from the African Development Bank argues that although the FEC claims to have instituted forty projects a year in this area, its training programme was insufficiently developed (BAD 2012). This makes it difficult to assert with certainty that the FEC contributes to improvements in demand for labour in the DRC.

Still, these may be significant initiatives for a post-war country that aspires to keep and attract investors. As several employers unlawfully and massively downsized their workforces during the transition, enlightening workers on such procedures might reduce increasing rates of job insecurity in the DRC.

Another illustration of overlapping laws that placed provincial and national state authorities in conflict relates to the issues raised by a piece of legislation called the Tax on Pollution (République Démocratique du Congo 2013).[7] This institutional conflict on the Tax on Pollution involved FEC's members, the national-level *Direction Générale des Impôts* (Directorate-General for Taxation, DGI), and the provincial-level *Direction Générale des Recettes de Kinshasa* (General Direction of Revenues of Kinshasa, DGRK) and affected employers who felt that they were overtaxed and fiscally harassed. Mr. Grace explained how the problem was resolved in 2016:

[7] The Tax on Pollution was instituted by the Ordinance-Law No. 13/001 of 23 February 2013 to Fix the Nomenclature of Taxes, Duties, and Fees of the Provinces and Decentralised Territorial Entities and their Distribution Modalities.

> The Congolese Business Federation, as the employers' trade union, continues to defend its members. ... [E]very year, the government publishes the state budget. And, the budget is a Finance Act. ... I take the case of the 2016 budget, in which there was a tax called the Tax on Pollution. ... It was said to be paid at the DGI [Directorate-General for Taxation]. The constitution stipulates that the provinces have rights to collect that tax. ... So both of these institutions harassed employers, each of them wanting to collect that tax. ... Talks were held with the government; we asked the government to make a decision. ... Then the government took a decision in favour of the province. You see how the Congolese Business Federation helped the government! (Mr. Grace, Lawyer, FEC, Personal Interview, Kinshasa, December 2016)

Like most labour market institutions in the DRC, the FEC encountered three main political and cultural challenges. During interviews, two informants criticised non-market actors for harassing and/or blackmailing employers. Mr. Grace, the lawyer, explained that,

> We have had several cases of the Congolese predation. So, there are certain persons in this country, just because they are ministers, or senior executives of the National Intelligence Agency or the army, they want to become shareholders in private companies without bringing their contributions to the capital of these companies. Shamelessly they threaten investors that they will ruin their business, expel them from the Congo, kill them, etc. (Mr. Grace, Lawyer, FEC, Personal Interview, Kinshasa, December 2016)

The economic difficulties related to over-taxation, poor infrastructure, inadequate foreign currency in banks, the dollarisation of the economy and negative impact of the global economy on local business. Mr. Robert, the FEC manager cited earlier, raised the issue of the informal importation of goods from Lufu, on the border with Angola[8] and its negative impact on local industry and the labour market,

> There is the problem of importing from Lufu, which perhaps makes consumers happy because they buy at low prices. ... [Yet] these imports have a negative impact on the local industry and the labour market. ...

[8] Lufu is located in Kongo Centrale Province (on the DRC's south-western border with Angola). Congolese and Angolans trade informally in this area. Congolese buy cheap but genuine goods imported from Brazil, Namibia, and South Africa and sell them at lower prices than they are sold in formal shops in Kinshasa.

> Some companies will close their doors; others will not see the light of day. (Manager, FEC, Personal Interview, Kinshasa, December 2016)

But he praised the FEC's contributions to the labour market:

> Without the effort made by Congolese Business Federation, who will talk about the Investment Code today? Without the effort made by Congolese Business Federation, who will speak about the National Agency for the Promotion of Investments, the one-stop shop? Without the effort made by the Congolese Business Federation, who will attract private investors, who create more jobs in the Congo? Without the efforts made by the Congolese Business Federation, who will protect employers better from the predators we have in the Congo? (Manager, FEC, Personal Interview, Kinshasa, December 2016)

Nonetheless, the FEC has been criticised for actually protecting unscrupulous employers, contributing to extreme precarity in the market and abusing government support. A trade unionist, for example, asserted that the FEC connived with employers who exploit workers:

> I would like to say that in the trade and supermarket sector, the Indians, the Lebanese, and the Chinese are bad employers. These people do not even like [there to be] unions. … But unfortunately, some politicians, senior military, agents of national intelligence, and especially the Congolese Business Federation are in collusion with these people. (Senior Executive, CGSA, Personal Interview, Kinshasa, December 2016)

Likewise, a retired unionist was of the view that FEC actions against employees put the labour market on near precarity:

> The Congolese Business Federation does not help our labour market, not at all. I cannot understand why it is opposed to any policy aimed at improving the wellbeing of the workers. Workers are big consumers and customers of the large corporations or big traders that the Congolese Business Federation protects blindly. … If the government does not control the Congolese Business Federation, our labour market will reach an extreme level of precarity. (Retired Trade Unionist #1, Personal Interview, Kinshasa, December 2016)

Additionally, members of the *24 entreprises et subséquentes* denounced the FEC in the media. Their spokesperson cited the inability of the highest institutions in the country to discipline the FEC by reporting its infringements to the United Nations Organisation Stabilisation Mission in the Democratic Republic of the Congo (MONUSCO),

> We do not understand how the Congolese Business Federation of Mr Yuma, its CEO, can prevent the execution of the decisions of the President of the Republic, the Prime Minister, the Minister of Portfolio, the Supreme Court, the Ministry of Justice, Ministry of Labour? How can Mr Yuma take all the institutions of our country as his hostages? This Yuma is the CEO of Gecamines today. Before that, he illegally dismissed more than 2,000 employees at the Utexafrica. But, why has the State done nothing against him?[9]

Some labour market actors see the FEC as protector of detested employers and that it exerts undue influence on the government. Some unions that operate in the informal sector denounce it for representing 'the interests of large companies, and thus mostly foreign interests' (Segatti 2015: 23). The FEC's influence on the government is also noted in the literature. Nkongolo-Bakenda et al., for instance, note the

> dispositional power of corporations operating in DRC [who] are members of [the] professional organisation called FEC (Federation of Enterprises of the Congo). Through this organisation, Congolese companies influence government decisions (Nkongolo-Bakenda et al. 2016: 479)

Similarly, Botedi (2013) notes that the FEC defends the interests of crooked employers who refuse to implement agreements on labour policies concluded between the government and unions. A report by the African Development Bank (BAD 2012) also confirms the FEC's power to mobilise various allies to lobby the government.

The FEC's contribution to the labour market, especially in terms of formulating and implementing labour law, reveals how employer organisations influence labour legislation in employers' interest. Similarly, the protection of employers against over-taxation and fiscal harassment, and the organisation of labour training, shows the tenacity with which it pursues the interests and rights of employers. As an employers' union,

[9] This speech is also available on YouTube (see Mazongelo, Aurelie. 2013. "Les Anciens Travailleurs De 24 Entreprises De La Rdc Reclament Leurs Salaires." *Congo Mikili News*. Retrieved 17 September 2017 [www.youtube.com/watch?v=pZLYSBnVZY8&t=180s]).

however, the FEC acts against the interests of employees. It impedes the implementation of various labour laws such as the minimal wage policy and the Supreme Court's decision in favour of members of *24 entreprises et subséquentes*. The FEC's extensive influence on the government can be understood through the notion of African neo-patrimonial states where patron-client networks strengthen political, economic and military supremacies (Bratton and Van de Walle 1994). Thus Moshonas (2018) observes that most heavyweight politicians in the DRC protect the interests of capitalists or are influential business people in their own right.

The close relations between the FEC and government are reflected in the fact that since 1981 the organisation (and its precursor, ANEZA) was chaired by only two men, both right-hand men of national presidents. Jeannot Bemba, businessman, politician, billionaire and close confidante of Mobutu, headed the organisation from 1981 to 1997. He was succeeded by politician and businessman Albert Yuma, chairman of Gecamines and a trusted and close confidante of Joseph Kabila (Lutete 2009). There are also reports of Yuma's deep involvement in corruption and the misappropriation of state funds (Carter Center 2017).

Employer organisations have shaped the labour market in the DRC from the colonial period to date. They use government and political authorities as allies to increasing their economic power by exploiting workers. Throughout the DRC's history the organisations have enjoyed privileges and accumulated wealth. It would, therefore, be adventurous to imagine the FEC as a positive contributor to the recovery of the labour market. Some may argue that it attracts investors and creates jobs and, thereby, contributes to the labour market. However, the foregoing evidence suggests that the institution's actions overwhelmingly focus on helping those who exploit the workforce. Because of this, its contribution leaves a lot to be desired.

Brief Historical Overview of Trade Unionism in the DRC

In considering trade unionism in the DRC, the five main types of trade unions according to Botedi are worth noting (2013). First, *syndicats du secteur privé* are unions that operate in the private sector. Second, *syndicats interprofessionnels* are private trade unions whose members are from interdependent professions. Third, *syndicats maisons* (company unions) or *syndicats jaunes* (yellow unions) are unions in which all members work for the same company (whether public or private) regardless of position. Fourth, *syndicats des employeurs* are employer unions or employer organisations. Finally, 'puppet unions' or 'empty-shell unions' are unrepresentative trade unions that are funded and used by government or employers in order to sabotage or suffocate labour activism by genuine trade unions.

Trade unionism in the DRC went through major stages of development in the time of the Belgian Congo. Although efforts by white employees to establish trade unions from 1914 onwards made the Belgian colonial authorities recognise the first white trade union in 1921, trade unionism in the country was strictly prohibited from 1908 to 1922 (Laurijssen 2010; Lwamba 1985). The period between 1940 and 1960 witnessed a second phase of trade unionism as workers pushed for legalisation. The activism pushed the Belgian colonial state to allow unions for Congolese in 1946 and non-racial unions in 1957 (Laurijssen 2010). The Belgian authorities nevertheless ensured that Congolese trade unions were under the control of the Belgian. A third phase of development occurred between 1960 and 1965 and was marked by a conflictual relationship between unions and political elites. While trade unions had strongly supported the anti-colonial struggle, the political elite now began to repress its former ally. In response, unionists sought aid from former enemies. They turned to foreign unions and Belgian state authorities in a quest to overthrow the political authorities (Mbili 1986).

The fourth phase, from 1967 to 1990, was defined by the nationalised *Union Nationale des Travailleurs du Zaïre* (Zairian National Union of Workers, UNTZA). The union was forced to incorporate itself into the ruling party and thus owed its loyalty to the party (Laurijssen 2010: 22). This placed it in a weak bargaining position vis-à-vis public and private employers (Botedi 2013). Though Mobutu's regime controlled this union, workers found measures to bypass the regime's force. In the late 1970s and early 1980s they began organising wildcat, sit-down and intermittent strikes (Young and Turner 1985). In response, the state established *syndicats maisons* or 'company unions'[10] to ensure close control of the workers and the workplace (Retired Trade Unionist #1, Personal Interview, Kinshasa, December 2016). In the fifth phase, from 1990 to 2002, the reinstatement of a multi-party system saw

[10] The Congolese French *syndicats maisons* means 'union of the house', thus trade unions where all members work for the same company (Botedi 2013).

trade unionism being allowed. Although a number of large trade unions emerged, they were less able to protect worker rights than in the 1990s because Laurent-Désiré Kabila was a ferocious oppressor of trade unionists (Botedi 2013). Up until the end of the war of 1998–2002, the DRC was thus not a friendly terrain for trade unionism. Trade unionism in the country has continued to evolve, nonetheless.

Trade Unionism from 2003 Onwards

Two main categories of unions exist in the DRC: one for the public sector and the other for the private sector. Separate legislation regulate the respective categories. Between 1981 and 2016 public sector unions fell under the Civil Service Department and were regulated by the Status of Civil Servants Act, No. 81/003 of 17 July 1981. Since 2016 the unions are regulated by the Status of Civil Servants Act, No. 16/013 of 15 July 2016. Private sector unions on the other hand fall under the Labour Department and are regulated by the Labour Code Act, No. 2002/015 of 16 October 2002 (Inaka 2020).

Added to this, two *nationales intersyndicales congolaises* (national platforms or inter-union organisations) exist in the DRC: one in the private sector and the other in the public sector. Both umbrella bodies, respectively, consist of 12 trade unions (Laurijssen 2010).

Three factors shape the contributions of unions to the post-war labour market in the DRC. First, the proliferation of unions has negatively impacted unionisation, employees' professional lives and households. Second, the 2002 Labour Code facilitated the involvement of unions in the informal sector. Third, unions can resist the hostile labour environments and help workers.

The proliferation of unions can be attributed to a range of political and organisational explanations. A retired trade unionist argued that the ruling party and private employers engaged in concerted efforts to divide and rule and, therefore, weaken the overall power of trade unionists. This strategy has been applied since Mobutu, when the DRC's Secret Service funded puppet unions to oppose the trade unions that were considered to be in cahoots with the political opposition. For them, it was essential to destroy unionists because they had gained political power. The situation in the 2020s might be worse than in the past, with numerous private employers supporting and creating 'yellow unions' (Retired Trade Unionist #2, Personal Interview, Kinshasa, December 2016).

A basic unionist view was that many considered creating a union as a business opportunity:

> Unions are created every day. We see these people as disguised unemployed ... it's actually a business proposition, that habit of creating yellow unions or puppet unions. Someone can just get up in the morning and create his union. I know someone who borrowed money somewhere, went to a company, ran his campaign, and told the workers: 'I am going to defend your rights'. And when he wins votes, he wants to recover what he spent on his campaign. Once in the company, the gentleman in question starts cronyism with the employer. You see! He deceived the workers for his own business. (Senior Trade Union Executive, Personal Interview, Kinshasa, December 2016)

The proliferation of unions weakens their functioning as the union leaders tend to focus only on their own economic and professional interests. A customer attendant at the supermarket chain store, Shoprite, explained:

> We don't trust these guys. ... They did not help us avoid becoming victims of outsourcing. ... They did everything to convince us to accept outsourcing. They were just spokespersons of Shoprite and outsourcers against our interests. ... They need only our money. We are waiting for the end of their mandate. It's better to be without a union instead of losing our money every month. (Customer Attendant, Shoprite, Personal Interview, Kinshasa, December 2016)

Many respondents emphasised this collusion between employers and yellow unionists and the way it affects the lives of employees during interviews. A manager at New Sarah, an Indian-owned supermarket, argued that unions fail to defend workers against low pay, excessive and long working hours, labour exploitation by outsourcers and poor working conditions. They even remained quiet when workers suffered occupational injuries and rather dismissed them instead of compensating them,

> This porter was carrying a huge wooden box. ... The guy slips, falls down, and that box injures him. ... It was horrible to see it. He was seriously crying, but Mr Yoga Gupta shouted: 'Give him US $50'. And he shouted: 'Go, I don't want to you see here again! I don't like lazy guys!' You understand? That's what I saw. (Manager, New Sarah, Personal Interview, Kinshasa, June 2017)

Despite the worker losing four fingers in the accident, the union did not help him receive an incapacity benefit:

> As a Congolese *mama*, I visited that young man. ... He lost four fingers. Imagine that, he is a right-handed person! Because of that accident, he can no longer work. You see! Despite an accident at work, he was fired, he didn't receive any compensation, and he lost his ability to earn. ... I realised that our unionists didn't care about him. One day, as the explanations of our heartless unionists about that man were getting on my nerves, I told them that they were evil, had sold out. I didn't insult them. I just reminded them what they are, for real! (Manager, New Sarah, Personal Interview, Kinshasa, June 2017)

The proliferation of yellow and puppet unions in the private sector opened up opportunities for employers to dominate unions and exploit their workers. A senior human resources manager at Airtel told interviewers how he personally experienced intimidation in his capacity as a leading unionist:

> I was the first president of the trade union in this company between 2000 and 2010. ... The Irish [former employers] understood us [unionists]. ... That position put me into trouble when the Indians came here in 2009. I was threatened, blackmailed, because I had defended some employees that the Indians wanted to downsize. ... These Indians were angry at me. I almost lost my job. ... To bypass me, they promoted me as a provincial manager and sent me to Lubumbashi, to be far away from Kinshasa, in 2012. After that, Indians completely crushed our trade union. ... Workers are just voiceless. (Maringa, Senior HR Executive, Yophone, Personal Interview, Kinshasa, June 2017)

Some scholars note that the proliferation of unions causes rivalries between trade unions because leaders seek to further their business aims instead of prioritising workers' rights and interests (Botedi 2013; Laurijssen 2010; Mpembele 2016). Such leaders consequently tend to accept bribes from employers they support and might even bribe comrades during union elections (IndustriAll Global Union 2016). Laurijssen (2010) argues that while the state in the DRC undermines unionists through the 'divide and rule' strategy, private employers use dual unionisation by pushing some leaders of trade unions to infiltrate rivals and create problems. Consequently, workers tend to lose trust in union leaders who are too weak to defend their rights. Botedi (2013: 61) summarised this thus, '*Trop de syndicats, pas de syndicats*' (Too many unions, no unions). In other words, a proliferation of unions has the same consequences as if no unions existed at all. The cited publications focused on the effects of the proliferation of trade unions in the workplace and not on how it affected employees' professional lives and personal lives, thus the aspect of labour reproduction (Peck 1996), which is a central aspect of the present study.

The trend towards recruiting new members from among informal workers can be seen as an adaptation to current dynamics of the labour market, which has seen the growth of the informal sector. A lawyer at New Sarah noted,

> Given the yearly expansion of our labour market in the informal sector, so that it has, in my view, become the most important labour market, integrating these people into unions is an adaption to current realities. ... What unionists do is important because they set up platforms which give voice to voiceless informal employees. (Lawyer, New Sarah, Personal Interview, Kinshasa, July 2017)

A retired unionist gave credit to that new trend towards formalisation of the unorganised labour market,

> It is important what they (the unions) do today. In our time, we didn't think about something like it. We looked for members in a few companies in the 1990s. Unionists can better stand for informal sector workers than NGOs. ... They understand that they are able to bring these workers' grievances to the attention of the government or employers. (Retired Trade Unionist #2, Personal Interview, Kinshasa, December 2016)

Similarly, the Vice-President of the Trade Union Confederation of Congo, Mrs. Joséphine Shimbi, acknowledged that the unions register members among informal employees due to the high rate of unemployment in the DRC. Speaking to Equal Times (2018), she stated,

> In 2012, only 2.5% of Congolese workers had what you can call 'formal' jobs — everybody else works in the informal sector. If we unionists say we are here to protect the workers, which workers are we talking about if only 2.5% have formal jobs? From there we started trying to see how we could really assist the country's informal workers.

One can agree with Mrs Shimbi and other informants that unions need to adapt to new trends in the labour market. However, using the high rate of unemployment as the sole rationale for union involvement in the informal labour market is also problematic. If unions already struggle to respond to the needs of the few existing formal employees, then they may face an overwhelming number of issues in the unorganised informal sector labour market.

Despite this misgiving, the extension of the 2002 Labour Code to the informal sector labour market might have greatly contributed to the new trend of integrating informal sector workers into unions. In this regard, Mrs Shimbi did not mention the guidelines of the new labour legislation on how unions may recruit members among informal sector workers. Segatti (2015), too, overlooked the impact of the 2002 Labour Code on that development. She found some positive effects when reviewing union roles in the informal sector in Kinshasa, such as unions fighting for the rights of workers in the informal retail industry. These unionists use mobilisations based on identity or autochthony over labour and Act No.73/009 of 5 January 1973, which reserved access to retail for Congolese, as their leitmotiv.

The proliferation of trade unions together with employers' opposition to unionism has led to resistance among members of the national inter-union movement in the private sector. The CGSA, for instance, denounced the fraudulent creation of 'alter ego employers' by some Lebanese, Indian, Chinese and Pakistani employers. These employers used deceit to illegally change the names and external characteristics of their companies, declaring their businesses as bankrupt but cunningly retaining ownership. A senior CGSA executive clarified this,

> I would like to say that Indian, Pakistanis, Lebanese, and Chinese in the retail sector are criminals. They do not like unions. … I personally fight against them. … But, unfortunately, some politicians, senior officers of the army, intelligence agents, and especially the Congolese Business Federation protect them. … When I investigated how they claimed the collapse of their businesses to the state, I realised that something was wrong. So, we investigated and found out that they were avoiding paying their employees money for termination of service. (Senior Executive, CGSA, Personal Interview, Kinshasa, December 2016)

Similarly, the national inter-union of the private sector organised actions against Chinese, Indian, Lebanese and Pakistani employers in the retail sector. According to Mabandu (2019), the inter-union sent a memorandum to President Felix Tshisekedi on 25 February 2019 to accuse Indian and Pakistani employers in the retail sector of conniving with state services to reduce the number of employees in order to avoid certain taxes and fees. In a company of 100 workers, for example, only 15 were declared and required to pay tax. The accomplices shared the money and saved on the 85 undeclared workers. As these employers also prevented trade union action, the inter-union also informed the president that most of the Chinese, Indian, Lebanese and Pakistani employers in the retail sector violated employees' labour rights. It followed that some Congolese who worked in Chinese, Indian, Lebanese and Pakistani retail shops demonstrated on 11–13 March 2019. They complained about the non-implementation of the minimal wage ordinance and the abuse of outsourcers in the retail sector. In response, Cabinet set up an ad hoc commission, composed of key members of the National Council of Labour and advisers of the president, and instructed them to resolve the issue of the minimal wage. As a result, members of the FEC persuaded the employers to implement Ordinance No.18/017 on minimal wages.

It is important to highlight that this action allowed the government to implement the same minimal wage policy that the FEC had vetoed less than a year earlier in May 2018. Two factors explain facilitation of the implementation of that policy in March 2019. First, the change of political authorities with the election of President Felix Tshisekedi in January 2019 softened the supremacy that the FEC previously enjoyed under Joseph Kabila. Second, trade unionists seized the window of opportunity to act collectively against employers by convincing the new political regime to decide in favour of workers. This solidarity made it possible to succeed in an environment of wider anti-union sentiments, especially among employers. It granted the unions a tactical advantage in an inimical setting, even if it was possibly only of temporary nature.

Unions have learned to respond to structural issues and adjust to the dynamics of the labour market by recruiting members from the informal sector. This adaptation has resulted in the formalisation of the informal sector labour market. As the vast majority of (informal) employees were less protected by law before the 2002 Labour Code was passed, it could be argued that trade unions contributed to the post-war reconstruction of the labour market in the DRC, and especially in the informal sector.

In sum, the proliferation of unions negatively impacted the post-war labour market because it prevented effective participation by unions. New labour legislation, on the other hand, has allowed unions to get involved in the labour market of the informal sector. This can be regarded as an adaptation by the unions to the dynamics of the labour market. Though the labour market generally remains opposed to trade unions, unions can better help workers when they unite to develop team strategies. On balance, however, the participation of unions in the labour market cannot yet claim a significant overall improvement.

In the description above, non-market actors are repeatedly identified as disruptive to labour market institutions. We now turn to analyse and evaluate the actions of non-market actors in relation to labour market institutions.

Non-market Actors

Non-market actors such as civil society and non-governmental organisations (NGOs), the DRC President's inner circle, religious bodies and influential individuals constitute a significant part of the labour market in the DRC. An examination of the country's labour market should, therefore, consider them. As labour markets are embedded in society (Peck 1996), non-market actors are, therefore, part of the labour market and cannot be separated from the study of labour markets. This makes it necessary to examine these actors and their actions in the labour market; to understand how the actions are sustained over time, and how the actors exercise their influence on the market and its reconstruction. A brief historical overview of the role of these actors in the labour market from the colonial period to the early twenty-first century helps shed light on their engagement with and in the DRC's labour market. We show how their practices have changed and adapted to changing situations and examine some arguments on the implications of their participation. We also consider how and why they remain non-compliant with labour legislation in their conduct.

Non-market Actors in Historical Perspective

Non-market actors have engaged in various activities including, violent resistance, hidden support for workers and collaboration with the state or private employers in the DRC's history. During the colonial period, traditional leaders and informal or formal labour brokers supplied workers to private employers and/or the colonial state (Lwamba 1985). They often used underhand methods, such as bribes and force, when recruiting Congolese workers in order to earn more (Reybrouck 2014). Employers and the state profited from this approach because it offered them cheaper labour. However, sometimes these forceful means met with violent resistance, and opposition from messianic movements. Marie Bolumbu, a female traditional healer, for example, waged war in 1914 against the abuses of the Belgian authorities. She was jailed from 1915 to 1929 (Hunt 2016). Another example is Simon Kimbangu, whose preaching of Congolese independence from the Belgian yoke in the late 1910s contributed to workers deserting workplaces, strike action and refusal to pay taxes. The colonial authorities imprisoned Kimbangu too in 1921 and he remained in their custody until his death in 1951 (MacGaffey 1982). The Bapende revolt in 1931 against the mistreatment of labourers by palm oil companies in Kwilu Province represented non-religious opposition in the labour market (Marchal 2017). Still, the colonial state disproportionately punished non-religious protestors too (Vanderstraeten 2001).

During the Mobutu era, non-market actors who actively engaged in the labour market included military officers, professors, politicians, Mobutu's siblings, and some religious leaders. Such figures exploited their political and socio-professional statuses within the regime to pressure employers to employ 'their people'. This did not change that much under President Laurent-Désiré Kabila and worsened during the transition period. Interference by non-market actors has since 2006 largely continued to affect the functioning of these institutions.

Non-market Actor Activities in the Labour Market

Non-market actors in the DRC give a wide range of reasons to legitimise their actions and activities in the labour market. These range from *jus soli* (native rights to land), to professional duty, labour discrimination, social justice, and to the high rate of unemployment. A traditional leader used the jus soli argument and argued that natives of Kinshasa must be prioritised on the Kinshasa labour market because of their right to the land. In response to an interview question, he argued thus,

> My son, do you know that all Kinshasa, in the centre where you stay, was our land? The state or other foreigners have taken it. We lost our villages … It is a matter of the owners of the land. It's not normal to see people from other places working here while Batéké children are jobless. They must be the first to work here. (Traditional Leader, Personal Interview, Kinshasa, April 2017)

An MP on the other hand saw it as part of his professional duty as member of parliament to help people find jobs:

> I am a member of parliament, my voters, comrades, neighbours, even my family rely on me. Why not to find something for them? It is my duty to find jobs for them. (Member of Parliament (A), Personal Interview, Kinshasa, July 2017)

On her part, a human rights activist spoke from a gender equality perspective to argue that labour market discrimination against women justified her actions:

> Women are victims of sexual harassment. … Look at all the governments that have been around since 2006: there were never 30% of women. This is the case everywhere. … We fight against sexism. (Human Rights Activist, Personal Interview, Kinshasa, February 2017)

A journalist was driven by altruism, to try and influence the labour market:

> As I told you, finding a job is a nightmare in the Congo. Everyone knows that. You see … we are all supposed to help people who are unemployed when we have the means to do so. (Journalist, Personal Interview, Kinshasa, March 2017)

Similarly, a professor remarked:

> It's difficult to become even a simple cleaner in the Congo. People are suffering to find work. Exactly, I feel pity for the unemployed. I have to help them as much as I can. (Professor, Personal Interview, Kinshasa, April 2017)

Religious motivation can also push the clergy to assist their members. A pastor disputed allegations that cultist employers who only hire members of their own groups flooded the labour market:

> In our country, especially in Kinshasa, there are many obstacles to access jobs. The Satanists [argue that] … to find a job, one must be or become a member of a sect like the Freemasons, the Rose Cross, the Maikari, the Lyons Club, etc. … All that pushes me to help my beloved ones. (Pentecostal Pastor, Personal Interview, Kinshasa, March 2017)

Our informants seemed to have one thing in common: a spirit of assisting others. Whether fighting for people's traditional rights, or gender equality in the labour market, or for voters, they all wanted to help particular groups of Congolese to get jobs or promotion at work. Their actions could be considered honourable because they addressed the exclusionary nature of the labour market. Assisting people who are socially disadvantaged because they are unemployed is commendable and, in that way, these actors contribute to reconstructing the labour market.

Yet a substantial difference exists between helping peers and observing labour recruitment regulations and procedures, such as entrance examinations, which are designed to ensure equity for all in the labour market. Respondents to interview questions for this book generally denounced the intervention by non-market actors in the recruitment, appointment and promotion of workers. Their general view was that, non-market actors encroach on the performance of administrative duties as set by decrees that govern the labour market institution for which they work. The intrusions suffocate the normal functioning of the institutions in all their aspects.

In terms of labour supply, statistics show that ONEM and the public service contributed little to recruitment procedures despite their legal mandate (INS 2014). This can be attributed to some non-market actors usurping the rights of the formal institutions. The state authorities that set up these institutions to resolve hiring process irregularities by pass them when they deploy non-market actors or practices in the labour market. This makes the labour market practices of the Mobutu regime era persist in post-war DRC.

The actions of non-market actors affected the management of human resources and industrial relations. Some workers refused to integrate into the authority structures of companies and organisations because they enjoyed the strong support of non-market actors (Kodi 2008). Foucault's (1975) notion of power as a constant process — an interaction, a struggle that is never resolved or 'balanced' —applies here. Non-market actors used their portion of power to influence the labour market and overstep its institutions. It is difficult to withstand the pressures of these persistent Mobutu regime era labour market practices in a country like the DRC, which is marked by neopatrimonialism. It is worth pointing out too that similar realities observed in public labour market institutions also existed in the private sector, as we show in the next chapter.

Conclusion

This chapter examined the contribution of the private sector to the reconstruction of the labour market in the DRC through Peck's (1996) view that the labour market is an arena of contestation, conflict, contradiction, complexity and paradox. The chapter engaged in a historical review of the behaviour of employer organisations and argued that since colonial times, the organisations encouraged the state to legislate and regulate the labour market in favour of capitalist interests. When in conflict with the state, employer organisations in the 1960s and early 1970s, for instance, manoeuvred to maintain economic privileges. On post-war reconstruction of the labour market, the chapter showed that the FECs always sought to satisfy its members through its actions. This institution had more privileges in the labour market because most political elites who ran businesses were its allies or members. Therefore, the FEC's contribution to the reconstruction of the post-war labour market in the DRC was limited by the undeviating capitalist pursuit of wealth, regardless of workers' rights.

The chapter further showed that trade unionisation in the DRC is marginalised throughout the history of the labour market. Trade unionism in the DRC emerged from workers' struggles against the colonial regime and employers. Trade unions struggled against different political regimes in the country. Immediately after independence, trade unionists became enemies of politicians, despite previously supporting the politicians in the fight against colonialism. The animosity of politicians towards trade unionists has not waned since. In the context of the post-war labour market, unions face challenges such as maltreatment orchestrated by politicians and employers, the negative effects of the proliferation of yellow unions, and the lack of trust of their members.

Despite the hostile environment, trade unions have managed to resist employers and the state and thus contributed to the democratisation of the labour market. Yet the hostile anti-unionist environment prevents them from making a sustained contribution to the remaking of labour market.

The chapter, lastly, identified the way that external actors presented constraints in the efforts to reconstruct the labour market. Those actors penetrated the labour market on grounds that they benevolently assist people at risk and yet are always narrowly defined in terms of their protégés. However compassionate or well-intentioned, the non-market actors often operated without due respect for or tactical awareness of labour market legislation. Their acts, therefore, affect the functioning and performance of labour market institutions and inhibit reform and renewal.

References

BAD, Banque Africaine de Développement. 2012. *Environnement De L'investissement Privé En République Démocratique Du Congo*. Tunis: BAD.

Bossekota, Leopold Watshia and Jacques Kiseta Sabiti. 2001. "Rd Congo: Dernier Pays Du Monde?" Vol. *Rapport d'OBSAM*. Kinshasa: Universite Technologique Bel Campus.

Botedi, Francois. 2013. *Carthographie Des Syndicats En République Démocratique Du Congo. Vers Une Comprehension Du Monde Syndical Congolais*. Yaounde: Presse Universitaire d'Afrique.

Bratton, Michael and Nicolas Van de Walle. 1994. "Neopatrimonial Regimes and Political Transitions in Africa." *World Politics* 46(4):453–89.

Buelens, Frans and Stefaan Marysse. 2009. "Returns on Investments During the Colonial Era: The Case of the Belgian Congo." *The Economic History Review*, 62:135–66.

Carter Center. 2017. *A State Affair: Privatizing Congo's Copper Sector*. Atlanta: The Carter Center

Depelchin, Jacquess. 1992. *From the Congo Free State to Zaire: 1885–1974: Towards a Demystification of Economic and Political History*. Dakar: Codersia.

Equal Times. 2018. "Congolese Trade Unionist Joséphine Shimbi Umba: "We Say 'Protect the Workers', but Which Workers Are We Talking About If Only 2.5 Per Cent Have Formal Jobs?"." Retrieved 9 October 2019 (https://www.equaltimes.org/congolese-trade-unionist-josephine?lang=en#.XOSqxfZuKJ8).

FEC, Fédération des entreprises congolaises. 2011. *Fédération Des Entreprises Du Congo. Status Revises Et Coordonees*. Kinshasa: FEC.

FEC, Fédération des entreprises congolaises. 2015. "Organigramme De La Fec: Restructuration De L'administration De La Fédération." *FEC*, Kinshasa

FEC. Retrieved 8 May 2019 (http://www.fec-rdc.com/index.php/notre-federation/organigramme).

FEC, Fédération des entreprises congolaises. 2019. "Nos Réalisations." *FEC*. Retrieved 9 May 2019 (http://www.fec-rdc.com/index.php/notre-federation/nos-realisations).

Foucault, Michel. 1975. *Surveiller Et Punir*. Paris: Gallimard.

Hunt, Nancy Rose. 2016. *A Nervous State: Violence, Remedies, and Reverie in Colonial Congo*. Durham and London: Duke University Press.

Inaka, Saint José. 2014. "Congolese Immigrant Workers in Pretoria, South Africa: A Sociological Approach in the Age of Migration." Dissertation, Master's thesis in Labour Studies and INdustrial Sociology, University of Pretoria, Pretoria.

Inaka, Saint José. 2020. "Post-War Labour Market Reconstruction: The Case of the Democratic Republic of the Congo." Doctoral dissertation, Sociology, University of Pretoria Pretoria.

IndustriAll Global Union. 2016. "New Union Challenges the Status Quo in Dr Congo." (http://www.industriall-union.org/challenging-the-status-quo-the-rise-of-tumec).

INS, Institut National de la Statistique. 2014. "Rapport Global, Enquete 1–2–3. Resultats De L'enquete Sur L'emploi, Le Secteur Informel Et Sur La Consommation Des Menages." Vol. *Rapport Global*. Kinshasa: INS.

Kahozi, Laurent Sumba. 2014. *Le Katanga Face Au Defi De L'inadequation Entre L'offre De Formation Et L'offre D'emploi*. Lubumbashi: Ministere Provincial de l'Education.

Kodi, Muzong. 2008. *Corruption and Governance in the Drc During the Transition Period (2003–2006)*. Pretoria: Institute for Security Studies.

Laurijssen, Eddy. 2010. *Pluralisme Syndical Et Prolifération Des Syndicats En Afrique Francophone*. Genève: BIT.

Lutete, Marcel. 2009. "'Congo-Kinshasa: Témoignage De La Fec Sur Le Président Jeannot Bemba Saolona'." Retrieved 1 December 2019 (https://fr.allafrica.com/stories/200907200785.html).

Lwamba, Bilonda. 1985. "Histoire Du Mouvement Ouvrier Au Congo Belge (1994–1960): Cas De La Province De Kattanga, These De Doctorat En Histoire." Faculte de Lettres, Departement de l'histoire, Universite de Lubumbashi, Lubumbashi.

Mabandu, Rachidi. 2019. "Exigeanr L'application Du Nouveau Smig, Les Employes Des Commercants Indopakistanais Radicalisent Leur Greve." In *Forum des As*. Kinshasa: Forum des As.

MacGaffey, Wyatt. 1982. "The Implantation of Kimbanguism in Kisangani, Zaire." *The Journal of African History* 23(3):381–94.

Marchal, Jules. 2017. *Lord Leverhulme's Ghosts: Colonial Exploitation in the Congo*. Verso Books.

Matenda, Athanase Kyelu. 2002. “Fédération Des Entreprises Du Congo.” *World Investment News* CBC. Retrieved 9 May, 2019 (www.winne.com/congo/to05.html).

Mazongelo, Aurelie. 2013. “Les Anciens Travailleurs De 24 Entreprises De La Rdc Reclament Leurs Salaires” *Congo Mikili News*. Retrieved 17 September 2017 (www.youtube.com/watch?v=pZLYSBnVZY8&t=180s).

Mbili, Mbili kwa. 1986. “Esquisse Du Syndicalism Au Zaire: D’hier a Aujourd’hui.” *ISIAO* 41(2):271–86.

Moshonas, Stylianos. 2018. *Power and Policy-Making in the Dr Congo: The Politics of Human Resource Management and Payroll Reform*. Brussels: Universiteit Antwerpen, Institute of Development Policy (IOB).

Mpembele, Guy Kisoka. 2016. “Promotion De La Démocratie Par Le Renforcement Du Plaidoyer Syndical.” Vol. *Rapport du consultant*. Kinshasa: intersyndicale Nationale du Congo.

Ndaywel, Isidore è Nziem. 1998. *Histoire Générale Du Congo. De L’héritage Ancien À La République Démocratique*. Bruxelles: Duculot Afrique Editions.

Ndaywel, Isidore è Nziem 2006. “Aux Origines De L’éveil Politique Au Congo Belge: Une Lecture Du Manifeste Conscience Africaine (1956) Cinquante Ans Après.” *Présence africaine* (173):127–44.

Nkongolo-Bakenda, Jean-Marie, Jeansy Kazadi, Robert B Anderson and Hilary Horan. 2016. “Privatisation, Stakeholder Power, and Weak Institutions: The Case of the Democratic Republic of Congo.” *World Review of Entrepreneurship, Management and Sustainable Development* 12(4):467–92.

Nzongola-Ntalaja, Georges. 2002. *The Congo: From Leopold Ii to Kabila. A People’s History*. London: Zed Book.

Peck, Jamie. 1996. *Work-Place: The Social Regulation of Labor Markets*. New York: Guilford Press.

Piret, Berengere. 2011. *Inventaire Des Archives De Félix Fuchs (1887–1927) Conservées Au Musée Royal De L’afrique Centrale*. Tervuren: Musée royal de l’Afrique centrale.

République Démocratique du Congo. 2013. *Ordonnance-Loi N° 13/001 Du 23 Février 2013 Fixant La Nomenclature Des Impôts, Droits, Taxes Et Redevances Des Provinces Et Des Entités Territoriales Décentralisées Ainsi Que Leurs Modalités De Répartition*. Kinshasa: Journal Officiel de la République Démocratique du Congo.

Reybrouck, David Van. 2014. *Congo. The Epic History of a People*. New York: Harper Collins.

Segatti, Aurelia. 2015. “‘Mobutu’s Ghost’: Mobilizing against Foreign Retailers in Contemporary Congo.” *Africa* 85(1):13–36.

Trefon, Theodore. 2010. “Administrative Obstacles to Reform in the Democratic Republic of Congo.” *International Review of Administrative Sciences* 76(4):702–22.

Vanderstraeten, Louis-François. 2001. "La Répression De La Révolte Des Pende Du Kwango En 1931." Doctorat en Sciences morales et politiques Sciences morales et politiques Bruxelles: Academie Royale des Sciences d 'Outre-Mer.

Vanthemsche, Guy. 2012. *Belgium and the Congo, 1885–1980*. Cambridge: Cambridge University Press.

Willame, Jean-Claude. 1995. "Entrepreneurs-Prédateurs Au Zaïre: Entre Le Grand Homme Et Les Affaires." Pp. 325–37 in *Entreprises Et Entrepreneurs Africains*, edited by S. Ellis and Y.-A. Fauré. Paris: Karthala.

Yates, Barbara A. 1978. "Shifting Goals of Industrial Education in the Congo 1878–1908." *African Studies Review* 21(1):33–48.

Yelengi, Nkasa. 2000. "Labor Policies and Household Economic Strategies among the Railway Workers in Katanga, Belgian Congo, 1928–1960." *Africa*:463–87.

Young, Crawford and Thomas Turner. 1985. *The Rise and Decline of the Zairian State*. Wisconsin and London: University of Wisconsin Press.

CHAPTER 7

Post-war Labour Market Reconstruction and Challenges in the Construction, Retail and Telecommunications Sectors

Job seekers in the Democratic Republic of the Congo (DRC) saw a continuous decline in opportunities or access to formal employment starting from the 1970s. The decline accelerated from the period of the Congolese war of 1998–2002. By the end of 2022, more than 80% of working-age Congolese were unemployed with only less than 3% holding jobs in the formal sector (Inaka 2020). A noteworthy observation in this regard is that labour market institutions on the labour supply side within the private sector in the DRC do not seem to collaborate with those on the demand side. Hence, the existence of formal employment in the sector appears to be miraculous. This calls for an examination of the ways that workers integrate into the private sector and the communication strategies used by potential employers to inform job seekers and the general public about available vacancies in their companies. It also calls for an analogous examination of the avenues that job seekers use to access information on available job opportunities. It makes sense to consider then, first, how workers get hired and deployed in various positions in companies such as those in the construction, retail and telecommunications sectors. For this purpose, Peck's (1996) notion of labour incorporation and Granovetter's (1973) theory of weak and strong ties provide useful tools with which to examine how job seekers in the DRC obtain information on job vacancies. Furthermore, Peck's (1996) notion of labour allocation

S. J. Inaka et al., *The Reconstruction of Post-War Labour Markets in The Southern African Development Community*,
https://doi.org/10.1007/978-3-031-37357-2_7

and Granovetter and Tilly's (1987) concepts of employers' recruitment networks and workers' supply networks also help to better assess hiring processes in the country.

Labour Incorporation

Peck (1996) views labour incorporation as processes through which individuals enter the labour market and have the potential to perform certain jobs while Granovetter (1973) proposes the concept of weak and strong ties to explain those processes. Granovetter argues that job seekers tend to use weak rather than strong ties in searching for jobs. Based on this, the question arose in this research whether individuals in the DRC use weak or strong ties when searching for jobs and information on vacancies. We proceeded to address this and related questions through an examination of the country's construction, retail and telecommunications sectors.

The Construction Sector

Figures 7.1 and 7.2 present information obtained from interviews with respondents who hoped to work in in the DRC's construction sector. The information included the positions for which job vacancies existed, the companies where those vacancies existed, the sources from which the job seekers obtained the information about the vacancies as well as the types of ties that existed between them and the suppliers of the information they received on available vacancies. In terms of organisational level and hierarchy, the vacancies in question existed at the levels of senior executive, supervisor and menial workers. The figures thus summarise the channels that job seekers used for information on vacancies in the construction sector of the DRC.

From Fig. 7.1, it is clear that available job vacancies in the construction sector tend to be communicated in four main ways including alumni networks, employers' recruitment networks (ERN), family members, friends and through labour recruitments that happen on the streets.

Overall, most of the respondents obtained information on job opportunities through family members while friends were least among the sources of information on job vacancies. In between were employers' recruitment networks, alumni networks and the streets. A further observation from the field and interviews was that, when considered in terms of the hierarchical levels of an organisation represented among

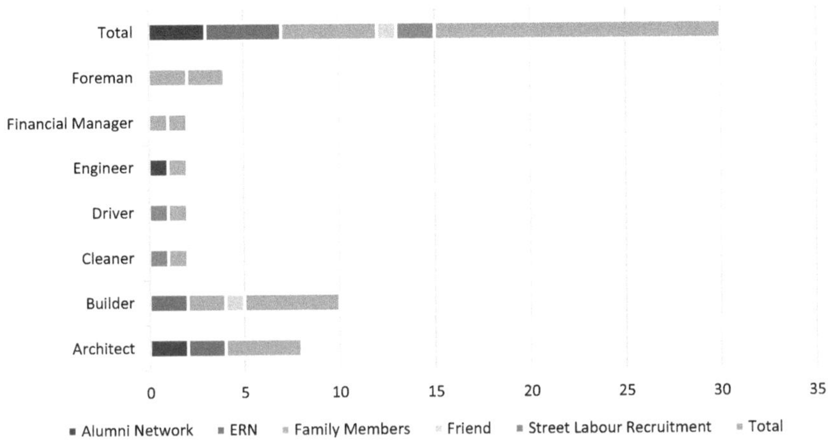

Fig. 7.1 Channels for information gathering processes on job opportunities: The construction sector (*Source* The authors. ERN: employers' recruitment networks)

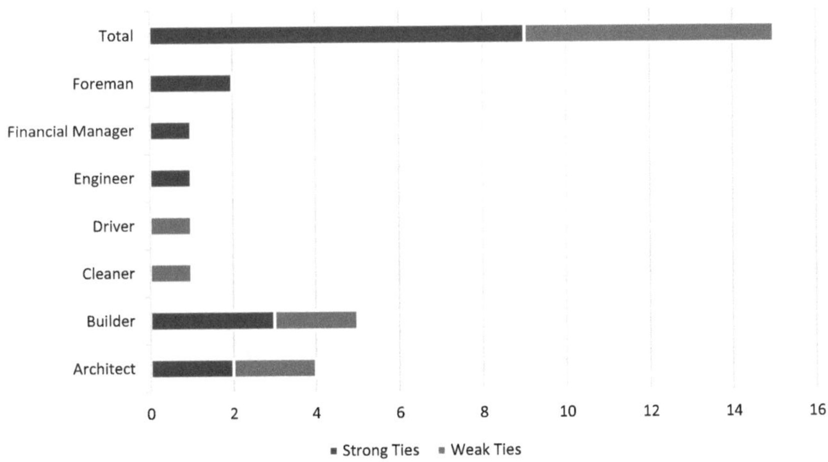

Fig. 7.2 Ties in information gathering processes on job opportunities: The construction sector (*Source* The authors)

the people who were interviewed for the study, alumni networks were the most common channel through which job seekers in the category of senior executive obtained information. For job seekers in the supervisors category, it was family members while for those in the menial worker category the channels of information were evenly distributed among family members, employers' recruitment networks and the streets.

Figure 7.2 also suggests that, strong ties defined more relationships in terms of the sources of information on job openings than weak ties. This was especially the case among job seekers of senior executive jobs and, to some extent, those who sought menial jobs.

Fieldwork data in this study showed that employers mainly provide information on vacancies, as they approach individual talented Congolese to offer them specific jobs. Sometimes this contact happens through other employers who convince talented engineers or architects to take up jobs on offer. The process is in this way initiated by employers and gives the potential candidate the power to negotiate wages and labour conditions. Kama accordingly shared his experience of being solicited by his employer thus,

> For me, it was not complicated because they found me on a building site where I was working. ... They explained to me their project. ... They asked me for a plan and quotation. I rapidly drafted it. ... They told me: 'We want to work with you'. (Kama, Architect, Globalium Business, Personal Interview, Kinshasa, April 2017)

Menial workers similarly seemed to benefit from networks between employers to access new jobs. Lutshatsha, a builder at KACA, narrated his experience:

> The boss observed us working at the house of another Indian. Then, he asked his brother to send us to him. ... We started working from then until today. (Lutshatsha, Builder, KACA, Personal Interview, Kinshasa, March 2017)

There was also a workers' supply network through which individuals accessed job opportunities. This happened when employers required employees (often architects and engineers) to source workers for their companies. Two architects, Mr. Basanga and Mr. Kadjeke, and an engineer, Mr. Bombi, explained that Congolese architects and engineers generally drew on their alumni networks, composed of former students of the *Institut des Bâtiments et des Travaux Publics* (Building and Public Works Institute, IBTP), for information on job openings (Basanga, Architect, KACA, Personal Interview, Kinshasa, March 2017; Kadjeke, Architect, Globalium Business, Personal Interview, Kinshasa, December 2016; Bombi, Engineer, KACA, Personal Interview, Kinshasa, April 2017). According to Mr. Kadjeke, the market for architects was saturated by people who lacked proper qualifications and sold their labour at cheaper rates. Foreign architects too flooded the market and tended to be brought from outside the country by their employers. But alumni networks enabled members to deal with this problem (Mr. Kadjeke, Architect, Globalium Business, Personal Interview, Kinshasa, December 2016). Mr. Basanga described his alumni network as a source of information on job openings in this way:

> When my colleague told me to replace him here (at KACA) because he had got public tenders to build stadiums, I couldn't hesitate to work with Indians as I was broke. (Mr. Basanga, Architect, KACA, Personal Interview, Kinshasa, March 2017)

The alumni network played two key roles in the labour market for architects and engineers. First, it helped employers find workers who possessed the required qualifications and skills. This implies that it protected employers from recruiting people who had fake qualifications. Second, it helped members get interesting jobs by blocking out non-members from information in circulation on job vacancies. The alumni network thus performed the role of an exclusionary professional enclave of former students of the same institution.

Recruitments dans la rue (recruitment in the street) provided information on vacancies to job seekers who gathered in certain public spaces in Kinshasa to offer their services, receive information on job opportunities shared between job seekers. Mr. Lukunga, a driver, described his experience as follows:

> As many people do it, I went to 13th Street in Limite where many people look for daily jobs. ... Someone who was passing told us that the Chinese needed five drivers. We were 11 persons to go to see these Chinese. (Mr. Lukunga, Driver, Ching Chong, Personal Interview, Kinshasa, March 2017)

Finally, the study found that none of the respondents interviewed within the construction sector used the services of ONEM to meet their job search needs. No employer sought ONEM's services either. A manager at KACA explained that they had once tried to recruit four workers from ONEM in 2013 but were disappointed because the institution sent them over 1,000 unsuitable candidates. They thus decided not to deal with ONEM any more (Mr. Elila, Manager, KACA, Kinshasa, Personal Interview, November 2016).

All informants in the construction sector employed unofficial methods to search for employment or to advertise vacancies. Though it is ONEM's responsibility to post job advertisements, private companies find it more effective to use their own unofficial practices to recruit employees. Two architects, Simonambi and Bulungu, thought that using unofficial channels to find candidates for jobs was explained by the absence of a national construction authority or a national order of architects. For that reason, they campaigned for the creation of the latter (Simonambi and Bulungu, Architects, Kinshasa, Personal Interview, February 2017). They did so for two reasons. First, they wanted to end the unfair advantage that foreign-owned companies held over Congolese companies because they bribed state authorities to accept their tenders. Foreign-owned companies also followed the local practice of using informal procedures for information on job vacancies. Second, Simonangi and Bulungu sought to curb the generalised disorder in the construction sector marked by violations of building norms, the hiring of bogus architects and unfair human resources management. Again they blamed the absence of a recognised professional body for this (Simonambi and Bulungu, Members of the Order of Architects, Personal Interview, Kinshasa, February 2017).

A regulatory body called the Order of Architects was eventually created on 13 December 2018 through the Creation, Organisation and Functioning of the National Order of Architects Law, No.18/034 (République Démocratique du Congo 2018). However, it was too early to assess and draw conclusions on the efficacy and efficiency of both the law and regulatory body in 2022, at the time of writing. Still, it is plausible to argue that the construction sector showed signs of becoming increasingly regularised than before the creation of this institution. Employer networks and strong ties among job seekers seem to have had a bigger impetus for its development than state-based labour market institutions. We observed similar patterns in the retail sector.

The Retail Sector

Figures 7.3 and 7.4 depict data provided by interview respondents regarding information gathering processes about job opportunities in the DRC's retail sector. The data, along with other responses to interview questions with job seekers suggest that while some of them used official channels to obtain information about job vacancies in the retail sector, many others used unofficial avenues.

Figure 7.3 shows that most of the respondents in this study got information about available job opportunities from friends and family members while very few made use of job search websites and employers' recruitment networks for such information. In between, were those who obtained information in situ, through outsourcers and from managers. When considered in terms of an organisation's hierarchical structure, most people who sought senior executive level jobs obtained information on vacancies through friends while information came in situ for those at supervisory, clerical and technical levels, respectively. Seekers of jobs that were menial largely relied on family members for information about vacancies.

In terms of ties, Fig. 7.4 indicates that more relationships in the processes of obtaining information about opportunities in the processes of

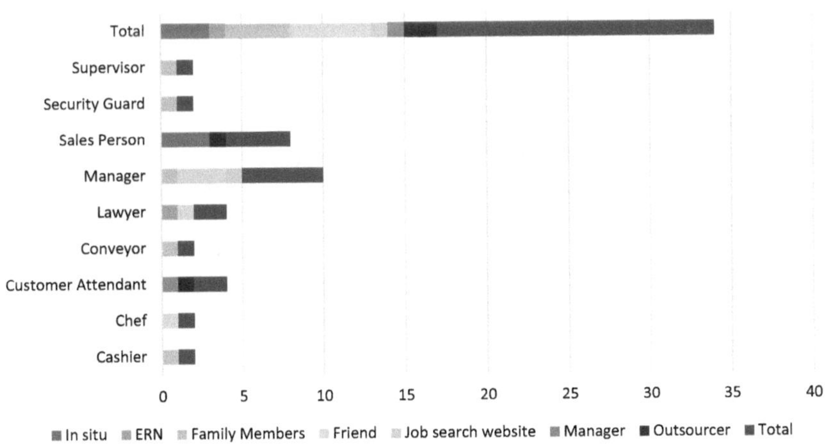

Fig. 7.3 Channels for information gathering processes on job opportunities: The retail sector (*Source* The authors. ERN: employers' recruitment networks)

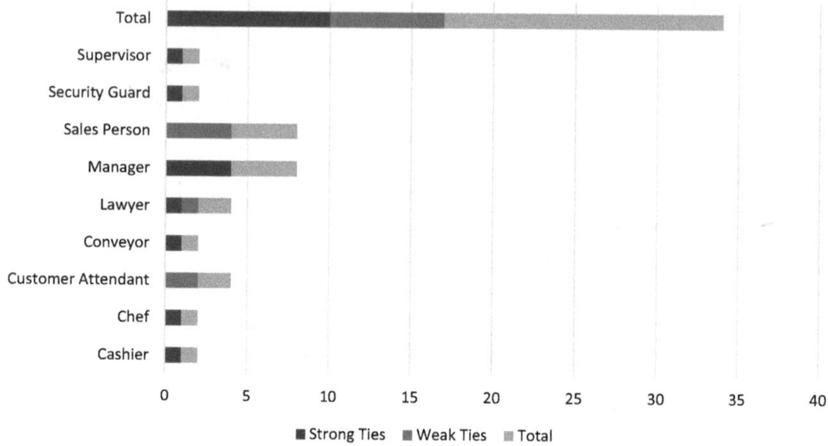

Fig. 7.4 Ties in information gathering processes on job opportunities: The retail sector (*Source* The authors)

searching for jobs were defined by strong ties than by weak ties. And from the perspective of organisational hierarchy or structure, more of the relationships through which information on vacancies was obtained among those who sought executive level jobs and menial jobs were defined by strong ties. However, the data also exceptionally revealed a mix of both formal and informal channels, both weak and strong ties, in processes of information gathering about job vacancies.

In 2009, Shoprite manager Mr. Kayange used formal methods of job searching and found job advertisements posted online on a job search website. He is of the view that he was positively influenced by the job-searching behaviour of South African students:

> When we were completing our B Comm at Vaal,[1] I saw many South Africans drafting their CVs, uploading them to websites, and calling in to follow up. I was, like, 'Why not do it like them?' ... I tried it several times. ... I received calls from three different recruiters. (Kayange, Manager, Shoprite, Personal Interview, Kinshasa, December 2017)

[1] 'Vaal' is a colloquial abbreviation for the Vaal University of Technology in South Africa.

Mr. Kwenge, a lawyer at New Sarah, simultaneously heard about a job vacancy from a close friend (strong tie) and the friend's employer (weak tie). In his own words,

> That day, God blessed me. ... My friend told me there was a (job) opportunity at New Sarah. He tried convincing me to go to see the boss. But I was reluctant ... It was (the boss's private assistant) who called to inform me that the boss needed me. (Kwenge, Lawyer, New Sarah, Personal Interview, Kinshasa, July 2017)

This narrative reveals how information on job opportunities can be channelled through both weak and strong ties. It also shows how the complexities of a network-based job search can move beyond Granovetter's (1973) dualism of weak versus strong ties. It is also revealing to note how employers take the initiative to inform talented people about job vacancies without following procedures for advertising jobs. In other words, information on job vacancies can originate from the demand side and flow directly to the supply side. Both Peck (1996) and Granovetter (1973) seem to overlook this pattern of information on job openings that is directly initiated on the demand side. Instead, they pay more attention on employees' initiatives in searching for a job (supply side).

Also noted in Fig. 7.3 is the use of outsourcers for gathering information about job vacancies. Lwange, a salesperson at New Sarah, explained:

> It is difficult to find a job here without having contacts. ... One day I went downtown, I saw a guy from an [outsourcing] agency. I approached him. ... I asked him to help me: 'My brother, I need a job whether it is here or elsewhere'. Well, he answered me: 'Prepare your file'. (Lwange, Salesperson, New Sarah, Personal Interview, Kinshasa, April 2017)

Concerning the in situ process, all four informants from Shoprite found out about job opportunities from their previous employers. As Lufira said:

> We were working as assistant builders. ... Shoprite people came from South Africa to find us here. They asked us to introduce our files and they hired us. (Lufira, Customer Attendant, Shoprite, Personal Interview, Kinshasa, December 2016)

It is important to note that this method of receiving information about job opportunities directly from employers occurs in unintended ways.

Instead of passing on job descriptions and clear specifications about what type of a person was required, and using conventional advertising channels, the Shoprite management randomly informed menial workers about available entry-level jobs. This method differs from the use of employers' recruitment networks. In the construction sector employers' recruitment networks are grounded in employers' capacities to take the initiative to look for talented people and detect them. In the retail sector, however, the Shoprite managers focused more on a prospective employee's observed working capacity than on screening their talents or training. A manager at Shoprite, Mr. Kasumu, noted:

> We recruit everybody with ability of working. I mean a person who is physically, mentally, and intellectually valid. That's all. We don't care about degrees or qualifications here. A worker learns everything that s/he does here in Shoprite. (Mr. Kasumu, Manager, Shoprite, Personal Interview, Kinshasa, December 2016)

In terms of labour legislation, these employers clearly did not follow Articles 203–207 of the Labour Code, which stipulate that employers must declare job vacancies at ONEM (République Démocratique du Congo 2002). In interviews, employers tended to react defensively when questions about these legal requirements were raised. There can be no doubt that they were well aware that they were not respecting them. The following response from Mr. Kibali at Shoprite puts it clearly:

> The Congolese employment policy exists on paper. These labour laws may seem interesting when you read them. In reality, the Congolese public authorities are the first ones who fail to strictly respect or implement the labour laws. ... The government or these authorities just want employers to pay their taxes for a National Employment Office. (Mr. Kibali, Manager, Shoprite, Personal Interview, Kinshasa, December 2016)

Overall, job seekers in the retail sector tended to find information on job opportunities through informal channels, drawing on strong ties rather than weak ones. Some employees, though, combined both strong and weak ties. It was indeed at times difficult to draw boundaries between weak and strong ties in this context because they intermesh in job-finding processes. It is thus plausible that in some contexts these ties are not mutually exclusive. There is also good reason to believe that the weak labour market institutions make the use of strong ties more effective than the legally prescribed procedures. This thus brings nuance to Granovetter's dualism between strong or weak ties.

The Telecommunications Sector

Besides sharing similar patterns of information flows on job openings with the construction and retail sectors, the telecommunications sector has an additional distinct attribute. As can be deduced from Figs. 7.5 and 7.6, transnational networks through which information on job vacancies flows in the telecommunications sector coexist with local networks. This can be attributed to the fact that the companies involved could not have used the services of ONEM because they existed before the organisation was established. Moreover, fieldwork data also points to ONEM's inability to meet the needs of these companies and, thereby, preventing collaboration.

As Fig. 7.5 shows, family members constituted the main channel through which information about job openings flows, followed by friends, return migrants' supply networks (RMSN) and marketers' supply networks (MSN). Hence, strong ties seem to be more important than

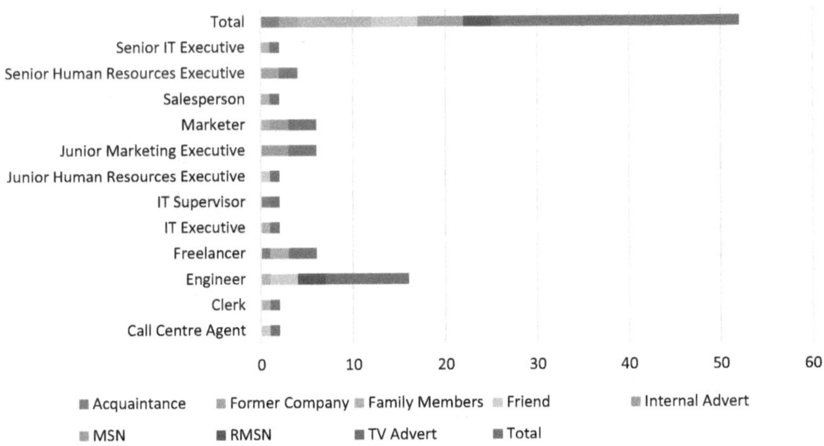

Fig. 7.5 Channels for information gathering processes on job opportunities: The telecommunications sector (*Source* The authors. CJSC: Conventional job search channels. MSN: Marketers' supply networks. RMSN: Return migrants' supply networks)

weak ties for transmitting information about job vacancies in the telecommunications sector. From the organisational hierarchy perspective, strong ties represented by family members as transmitters of information were particularly noticeable with Routine Clerical, Supervisory, and Technical Workers as well as Operational Delivery Workers' jobs. For Junior Executive level kind of jobs, most of the information was transmitted through friends.

Figure 7.5 also shows two additional networks, seemingly not catering to job seekers in the construction and retail sectors. The RMSN and MSN constituted two additional and specific means through which information on job openings in the telecommunications sector was obtained. RMSN were unstructured networks of engineers in the telecommunications sector who had graduated from South African institutions and provided their Congolese counterparts with information on job openings in the late 1990s and earlier 2000s.

During interviews, two engineers, Chris Kwa and Dany, explained that most senior managers in telecommunications companies in the country were South Africans. The managers preferred hiring Congolese engineers who had graduated from South African universities to those from local universities (Chris Kwa and Dany, Engineers, Yugphone, Personal Interviews, Kinshasa, January 2017). The networks played a significant role in appointment processes as the companies allowed engineers to find other Congolese with similar training whenever vacancies occurred. Thus, engineer Dany was introduced to Yugphone by a former colleague from the Durban University of Technology. When he in turn was asked to find an engineer with similar qualifications, he suggested a colleague who had trained at Tshwane University of Technology (Dany, Engineer, Yugphone, Personal Interview, Kinshasa, January 2017).

The reason South African employers preferred Congolese engineers who had trained in South Africa was because those trained in the DRC had poor skills. Another reason had to do with communication challenges between the English-speaking employers and French-speaking employees. Gomes et al. (2013) indicate that because the skills of engineers and technicians trained in the DRC did not match the competencies required by companies such as Vodacom, DRC-trained engineers and technicians 'were sent to South Africa to attend training and courses provided by the Vodacom Group to learn about Vodacom processes and the use of new technology' in the late 1990s and earlier 2000s. Return migrants' supply networks informed employers about candidates who had tailor-made skills needed by their companies.

Another workers' supply network in telecommunications involved marketers. Information about job openings at telecommunications companies was circulated through this network. Mr. Maheshe, a marketer himself, argued that it was the need for marketers in the early 2000s that made companies use these networks to circulate information about job vacancies:

> I know how to convince customers, to persuade someone to consume products. ... I know well how to resist to that. But they (other marketers in his network) showed me evidence that pushed me to leave British American Tobacco. (Mr. Maheshe, Junior Marketing Executive, Yugphone, Personal Interview, Kinshasa, January 2017)

As Fig. 7.5 shows, job seekers in the category of Senior Human Resources Executives obtained information on job vacancies from companies they had previously worked for. The particular respondents were actually transferred to their current companies due to the change in ownership of the company in the late 1990s. The transferring of employees to new companies is telling, because it shows how employees might become members of new companies without the need to seek information on available positions. Moreover, it adds formally registered telecommunications companies to the list of organisations where strong ties define information gathering processes about job opportunities.

And, as Fig. 7.6 shows, strong ties even define information sharing processes on vacancies among professionals such as engineers in a specialised sector like telecommunications. Strong ties contribute to the recruitment of such professionals. Hence, Chris Kwa and Dany's experiences and their disclosure during interviews confirm our observation that strong ties too can characterise information gathering processes about job vacancies among professionals and that this could even be influenced by the preferences of employers.

The processes through which job vacancies were advertised in the telecommunications sector also varied across companies and job positions. Yugphone, for example, first advertised vacancies internally and only placed the job adverts on its website if no internal candidates responded to them (Mr. Okapi, Engineer, Yugphone, Personal Interview, Kinshasa, January 2017). At Wangphone, the Chinese executives

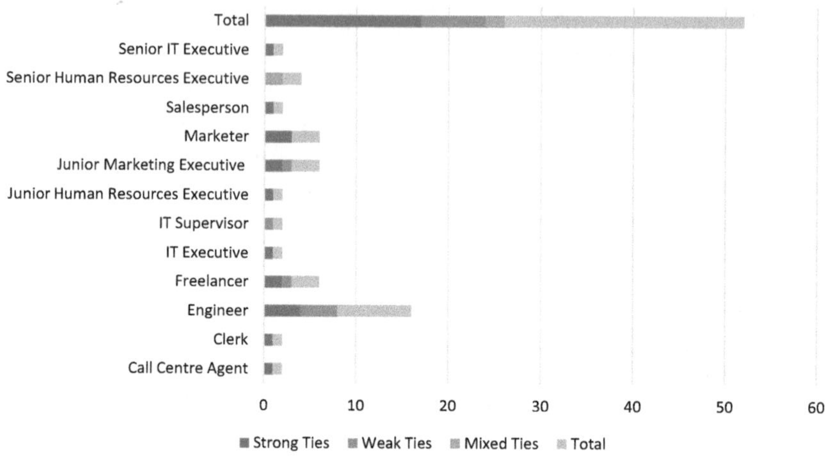

Fig. 7.6 Ties in information gathering processes on job opportunities: The communications sector (*Source* The authors)

used verbal communication as strategy for secretly selecting candidates, even before internally advertising the jobs. In the opinion of one of its engineers, Mr. Rubi, these executives did not value ONEM's services. They rather paid the penalties for failing to declare their job vacancies at that office (Mr. Rubi, Engineer, Wangphone, Personal Interview, Kinshasa, January 2017). In contrast, Mzansiphone simultaneously informed everyone about vacancies, despite prioritising internal placements. Mzansiphone also informed ONEM about available vacancies but the latter often failed to perform its procedural role in identifying possible candidates (Mr. Tuana, Senior Human Resources Executive, Mzansiphone, Personal Interview, Kinshasa, July 2017).

It is worth noting that these three companies, Yugphone, Wangphone and Mzansiphone, existed before ONEM was established. They might, therefore, have respected the regulations that existed at that time and it might be unwarranted to judge their using illegal channels to communicate job vacancies. Even if it has not lived up to its task, ONEM only became functional in 2012. Still, the foregoing indicates ONEM's inefficiency and poor performance and that this negatively impacts employers and job seekers.

The research also interestingly observed mixed ties (Fig. 7.6) and the possible effect of those ties on information gathering regarding job search processes and possible consequences for post-war labour market reconstruction. Mixed ties occurred when a single job seeker used both strong ties and weak ties to gather information on job vacancies. To us, this suggested that a 'Chinese wall' did not exist between strong and weak ties. It also meant that ties can overlap or not during a job seekers' labour incorporation process.

Labour Incorporation

A critical issue in the labour market of the DRC is the manner in which job vacancies in the private sector are communicated. Weak labour regulations, weak key official organisations that are mandated to provide information on vacancies, and the scarcity of critical skills or job opportunities reinforce the use of unofficial methods to search for jobs.

The prolonged absence of institutions to regulate the relationship between employers and employees in the hiring processes is critical. The Mobutu regime created SENEM in 1967 to deal with issues concerning this relationship. SENEM was never operationalised, however, and private companies came up with their own means of finding and hiring workers. After the 1998–2002 war, ONEM was created to fulfil the same objectives as its predecessor, SENEM. But it too faces problems that relate to the political, administrative and sociocultural environment in the labour market and struggles to meet the expectations of both job seekers and employers.

Job seekers, therefore, use both weak and strong ties in the labour incorporation process and this proves different from Granovetter's (1973) argument that a dualism exists between weak and strong ties. The market significantly uses both weak and strong ties simultaneously in certain circumstances. Employers tend not to follow labour legislation but rather draw on employers' recruitment networks and workers' supply networks to establish bridges between employers and employees. Thus, it is a challenge to obtain information on job vacancies through legal and regulated channels in this labour market.

Hiring Processes and Labour Allocation

Informal practices exist in all sectors of the labour market covered in this study. However, differences exist from one industry to another. While employees in the retail and construction sectors are often hired by 'illegal' outsourcers or through employers' or workers' supply networks, those in the telecommunications sector tend to use formal hiring processes that are directly organised by companies. The construction sector is also marked by informal hiring processes such as verbal agreements between employers and employees. This prompts a discussion of labour allocation in these three sectors as follows.

The Construction Sector

Analysis of informants' responses to interview questions points to the existence of two procedures for hiring workers in the construction sector. The procedures also vary according to job position for which job seekers are vying and the company that is hiring. One of these procedures included negotiating fixed-term employment contracts between employer and employee. This process tended to be used in hiring senior executives of companies. Another involved informal hiring and was used in hiring workers at both the supervisory and menial levels. Figure 7.7 depicts these hiring processes for various job positions in the construction sector.

Overall, Fig. 7.7 shows that most of the informants in the study negotiated fixed-term employment contracts, followed by those who got part-time employment contracts and those who got contracted through verbal agreements. Of the 15 people interviewed in this regard, a female worker, Mrs Elila, a financial manager at KACA, was the only person who got hired through traditional hiring processes. Her contract was a pre-formulated standard contract whose contents and clauses were unilaterally drafted by the employers (Mrs Elila, Finance Manager, KACA, Personal Interview, Kinshasa, March 2017).

The engineers and architects all confirmed that their jobs corresponded with their professional and academic profiles. Despite that, prospects of being hired as full-time permanent employees were uncertain. As Basanga reported:

> We agreed on my remuneration and labour conditions. But the employer refused to offer a full-time permanent employment contract. ... I had

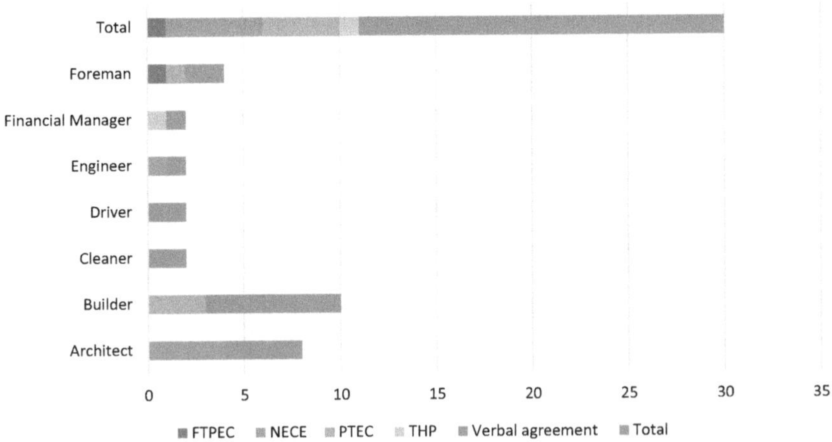

Fig. 7.7 Hiring processes in the construction sector (*Source* The authors. THP: Traditional hiring process. THP involves choosing, short-listing and interviewing the best job candidates, recruiting them, orienting or training them, and co-signing an employment contract [Fernández-Aráoz 2001]. NECE: Negotiation of fixed-term employment contracts. FTPEC: Full-time permanent employment contracts. PTEC: Part-time employment contracts)

> no choice but to accept the status of a part-time employee. (Basanga, Architect, KACA, Personal Interview, Kinshasa, March 2017)

Kama, an architect, was familiar with this trend towards part-time employment in construction. He was thus very careful when negotiating his wage and working hours:

> As they (employer and his staff) said they wanted to work with me, I had an opportunity to raise the bar higher. ... I was very strict with respect to my working hours. (Kama, Architect, Globalium Business, Personal Interview, Kinshasa, April 2017)

Two architects, Mr. Simonambi and Mr. Bulungu, helped explain a common practice in the DRC of hiring architects and engineers on temporary bases. According to Simonambi, very few chances of securing permanent employment existed for engineers and architects in the labour market of Kinshasa. For him, this is because most of the European-owned companies, which used to offer these positions to Congolese employees, had shut down since the lootings of the 1990s (Simonambi, Order of Architects, Personal Interview, Kinshasa, February 2017). Bulungu added that many companies often temporarily hired architects to avoid paying architects in the absence of job opportunities. Generally, therefore, employment contracts between architects and their employers only run to the end of a building project (Bulungu, Order of Architects, Personal Interview, Kinshasa, February 2017).

It is worth noting that some employers hire informal employees through conventional hiring procedures—selecting, screening, testing job candidates and hiring the best (Stewart 2016). However, people who were mostly hired in this way had no choice but to accept underemployment and downgraded job positions or remain unemployed and poor. Lukunga, a trained computer scientist, was hired as a driver and described the pressures he experienced during the hiring process:

> I was the only one who had a driving license. … They tested and accepted me. … They told me I would get U$5 per day. … No, there was no (written) contract. … Between us, let's be clear, there is no match between a well-trained computer scientist and a driver for Chinese builders. (Lukunga, Driver, Ching Chong, Personal Interview, Kinshasa, March 2017)

Although such employers followed conventional procedures of hiring employees, they failed to respect the legally encoded hiring procedures: 'hiring of a worker shall be subjected to a contract in writing and concluded for a fixed or indefinite (National Agency for the Promotion for Investment 2016). Here it seems inappropriate to argue that these employers preferred hiring Congolese through informal processes. As argued earlier, the National Order of Architects that was only created in December 2018 (Ordre National des Architectes 2019) and the weakness of ONEM prevent the construction sector from being effectively regulated.

In such a context of institutional weakness, one might expect employers to use hiring procedures they consider best suited for their companies. Since each company seems free to follow any hiring procedure, it is not inconsistent to consider that labour market institutions have a large share of responsibility for these disorders. Theoretically, we have seen that the role of labour market institutions is to ensure harmonious regulation of the labour market (Berg 2015). But this principle does not apply here. The hiring procedures described above are so deeply anchored in the practices of the construction industry that it is difficult to change them through labour legislation.

The Retail Sector

In response to interview questions, employees of Shoprite and New Sarah indicated that companies in the retail sector hired workers through formal procedures. This can be seen in Fig. 7.8, which depicts information about hiring processes in the retail sector. Figure 7.8, however, also shows that outsourcers have a strong influence on hiring processes in this sector.

Figure 7.8 also shows that most of the workers in the retail sector interviewed for this book went through four different processes of hiring. The first process involved the transnational recruitment of managers, wherein multinational companies or transnational entrepreneurs in the DRC hired Congolese who had worked or studied overseas and then transferred them to their operations in the DRC (see Pruthi and Wright 2017). All managers at Shoprite in Kinshasa had gone through such processes of transnational recruitment. All of them had graduated from South African universities. Shoprite hired and sent them to open and manage its branch in Kinshasa (Kayange, Kibali, and Kayange, Managers, Shoprite, Personal

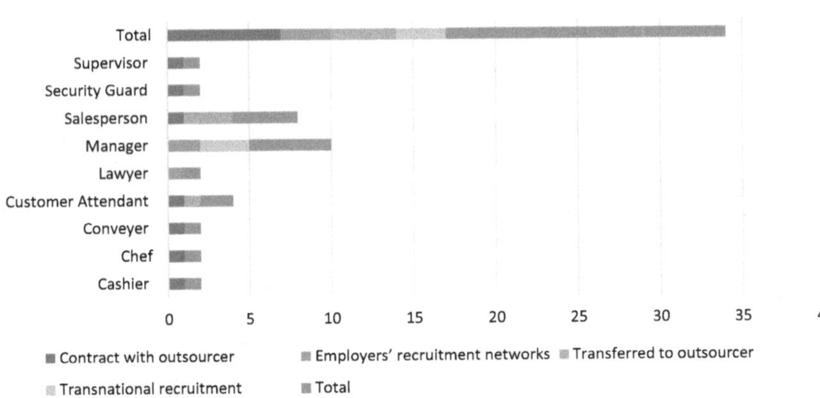

Fig. 7.8 Hiring processes in the retail sector (*Source* The authors)

Interview, Kinshasa, December 2016). Mr. Kayange described the hiring process as follows:

> We studied in South Africa. All of us. Like, Ligbafu was at the University of Johannesburg, I was at the University of Cape Town. … After three months of training (at Shoprite) in Cape Town … we came to open here in 2012. … They just wanted Congolese to work for them in Congo. Ultimately, it was a well-conceived strategy to send us here. (Kayange, Manager, Shoprite, Personal Interview, Kinshasa, December 2016)

Unlike all other foreign-owned companies in this study, Shoprite (a South African company registered on the Johannesburg Stock Exchange with a presence in a growing number of African countries) has developed an exceptional labour allocation marked by the transnational transfer of managers, whereby all its managers in the DRC are Congolese. Data from interviews for this study showed that the managers hired 156 Congolese on full-time permanent bases through transnational transfers from 2012 to 2014. This seems exceptional because the literature (Muller-Camen et al. 2001; Pruthi and Wright 2017) does not show any other cases of a multinational company that transfers natives only, as does Shoprite for Congolese, to manage its branches in the employees' country of origin.

The second hiring process of interest here drew on employers' recruitment networks. The intermeshing of strong social ties and elite employers' recruitment networks in the DRC facilitated the hiring on permanent bases of some respondents as senior executives. This hiring process involved a degree of *quid quo pro* between some elite Congolese and foreign employers. For instance, Kikimi indicated how mutual trust between some elites and Indian employers' recruitment networks helped her get a position at New Sarah. Kikimi's story shows how using strong and weak ties can simultaneously influence both labour recruitment and hiring processes:

> I studied in India. Okay! When I was in New Delhi, we had our small community. … Well, after my studies, I returned to Congo. … As it happened, while I was in my former job, a friend — the ex-adviser of that New Delhi community — came to visit me. … It was he who introduced me to my new boss, as his father, Mr. Mangbanga, is associated with the Indian business community. (Kikimi, Manager, New Sarah, Personal Interview, Kinshasa, June 2017)

Another manager at New Sarah, Mr. Miao, son of a prominent Congolese politician, informed the interviewer how his father's influence played a key role in his hiring.

> When I came back from India…my father placed me at BIAC.[2] Unfortunately, BIAC had problems. So I was again unemployed for three months. … One day, my father told me that there was just the right position for me at New Sarah. … The following day, the boss sent his driver to fetch me to start working. (Miao, Manager, New Sarah, Personal Interview, Kinshasa, March 2017)

The respondents' narrations above show how strong and weak ties can work simultaneously during hiring processes. They also point to the continuing encroachment of non-market actors (like politicians), with its corollaries in the remaking of the labour market of the DRC. The responses validate how unlawful hiring practices in the public sector are reflected at the company level where patron-client networks facilitate these processes.

In a third processes of hiring, informants signed employment contracts with outsourcers. Some outsourcers obliged job candidates to pay fees for job applications. This echoes an ONEM report (ONEM 2015) which denounced private job placement services for taking fees from job candidates. On top of that, most of the respondents to interviews for this book felt forced to agree to being underemployed. Idiba, a graduate in International Relations assigned by an outsourcer to the post of salesman at New Sarah argued that the practice of requiring job candidates to pay a fee was an 'organised swindle':

> He (the outsourcer) asked me to pay US$10 for application fees. It's the requirement of the agency. I took a week's training, but I didn't start working at that time. They told me I failed in the training. I had to redo the same process (meaning to pay another application fee and get trained again). … That's how they perfectly scam poor jobless people. (Idiba, Salesperson, New Sarah, Personal Interview, Kinshasa, April 2017)

[2] *La Banque International de l'Afrique Centrale* (International Bank of Central Africa).

The fourth process was the forced transfer of permanent employees into temporary outsourced employees. Four informants at Shoprite experienced this. In 2012 they signed up as full-time, permanent employees, only to find out that their employer had transferred them to an outsourcer in 2014. This outsourcer required them to renew their employment contracts every three months, deducted 10% from their monthly salaries, and was quick to fire employees (Lufira, Customer Attendant, Shoprite, Personal Interview, Kinshasa, December 2016; Lubefu, Lumene, and Lua, Salespersons, Shoprite, Group Discussion, Kinshasa, January 2017). Lua recounted what happened:

> When we started in 2012 there was no outsourcing. ... They [Shoprite managers] introduced an outsourcer in a malicious way. ... In reality, outsourcers come with their employees. But here, he (the outsourcer) came alone. (Lua, Salesperson, Shoprite, Focus Group Discussion, Kinshasa, January 2017)

According to Shoprite managers, Kasumu and Kayange, the state authorities in the DRC forced them to transfer full-time permanent employees to outsourcers. In fact, Kasumu argued,

> we wanted to treat employees as they do it in South Africa. Unfortunately, we faced pressure from everywhere. (Kasumu, Manager, Shoprite, Personal Interview, Kinshasa, December 2016)

Kayange added:

> We hired them on a permanent basis when we started in 2012, but we were forced by authorities of the National Intelligence Agency, Generals of the Army, Ministers' advisers to outsource them. (Kayange, Manager, Shoprite, Personal Interview, Kinshasa, December 2016)

Although assertions of this pressure by the authorities on outsourcing might be credible, Kasumu's argument that hiring processes in Shoprite in the DRC maintained the same standards as those of Shoprite in South Africa—based on full-time permanent employment contracts—is questionable. In fact, there is evidence that Shoprite widely used outsourced workers in several of its locations: South Africa (Webster 2005), Namibia (Mathekga and Maciko 2018) and Nigeria (Akanji 2016). Some journalists (Goko 2016; Maromo 2016) also cite Shoprite among key targets of the 2016 South African 'OutsourcingMustFall' movement, which provoked several strikes against the outsourcing practices in South Africa.

A close look at labour-hiring processes through outsourcers in the retail sector confirms that they operated in this way before the Labour Act No. 17/001 of 8 February 2017 on Outsourcing in the Congo (République démocratique du Congo 2017) was passed. Instead of acting as private intermediaries between employers and employees, private job placement services worked as outsourcers. This is among the reasons why Labour Minister Bahati chastised many private job placement services in 2015. There is no doubt, thus, that hiring processes that occurred through outsourcers before the promulgation of the 2017 Labour Act had no legal grounding.

Overall, transitional labour allocation, mixing strong and weak ties, illegal outsourcing, and the transfer of permanent employees to outsourcers constituted the main hiring processes in the retail sector. While transitional labour allocation and the use of a mixture of strong and weak ties facilitated the hiring of informants as managers, outsourcing led employees into unfair hiring processes. Such practices existed in the telecommunications sector too.

The Telecommunications Sector

Responses to interview questions on hiring in the telecommunications sector indicate a correlation between information gathering on job opportunities and patterns of labour allocation in the sector. This is evident in Fig. 7.9. Job seekers who obtained information about job opportunities through conventional channels were more likely to get permanent jobs through labour contracts directly signed with employers. Conversely, job seekers who obtained information on jobs through unofficial channels were more likely to be outsourced or hired as temporary employees.

Figure 7.9 also clearly shows that job seekers who aimed for executive level positions tended to secure permanent positions contrary to jobs whose functions were routinely clerical, supervisory, technical and operational.

Figure 7.10 shows six main processes through which most of the respondents in the study for this book got hired. The first process involved direct hiring by outsourcers. The second saw job candidates and employers following conventional hiring procedures. The third involved permanent employees being transferred to outsourcers. The

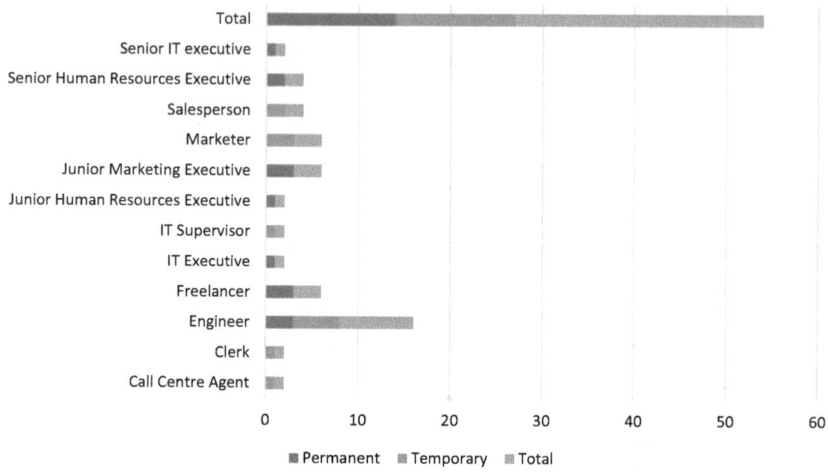

Fig. 7.9 Types of jobs secured in the telecommunications sector (*Source* The authors)

fourth involved companies hiring students and granting them scholarships to study at a South African university. The fifth involved managers being transferred to a new company that was a joint venture or had alliances with the former company. Lastly, companies used campus recruitments, where they visited university campuses within the DRC to select students and hire them to join the companies upon graduation.

Of the six cited hiring processes in the telecommunications sector, this research found the practice of transferring engineers to outsourcers particularly intriguing. Situations occurred where outsourcers deceived employees that they were hired on long-term labour contracts in given telecommunications companies. A previous section in this chapter has, however, shown that in the late 1990s and early 2000s, senior South African telecommunications managers preferred hiring Congolese engineers who had received training in South Africa. At that time, the engineers were hired as full-time permanent employees. However, respondents disclosed that they ended up being transferred to outsourcers. This happened for two interrelated reasons. Two engineers who were interviewed for this book indicated that rapid technological change and especially the introduction of automation in their company contributed to the transfers in the early 2010s. The engineers had no choice in

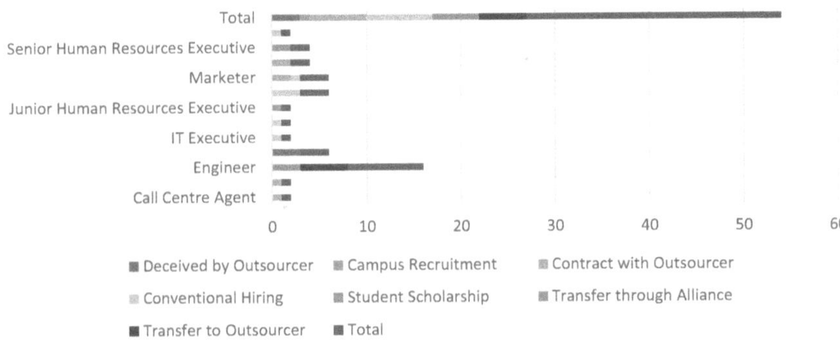

Fig. 7.10 Hiring processes in the telecommunications sector (*Source* The authors)

the matter than to accept becoming casual workers with outsourcers (Dany and Okapi, Engineers, Yugphone, Personal Interview, Kinshasa, January 2017). Because of the shortage of engineers in the telecommunications sector in the early 2000s, Dany, an engineer, had negotiated and managed to get himself a well-paying job. After 2010, the Chinese company Huawei brought cheaper sophisticated technologies into the DRC and forced several other telecommunications companies such as Erickson, Siemens and Alcatel out of the market. The telecommunications labour market simultaneously got saturated with technicians and engineers, trained in other parts of the world. This led to full-time permanently employed engineers either getting transferred to outsourcers or losing their jobs. Dany noted regretfully that many engineers were subsequently underpaid, underemployed and vulnerable to the outsourcing sector (Dany, Engineer, Yugphone, Personal Interview, Kinshasa, January 2017).

During interviews, Mr. Maringa, a senior Human Recourses Executive at Yugphone, disclosed that his company often outsourced and downsized because of declining profits, international financial challenges and changes in company ownership (Maringa, senior Human Resources Executive, Yugphone, Personal Interview, Kinshasa, June 2017).

Another engineer at Wangphone, Mongala, disclosed that recurring changes in ownership of his company led to the departure of South African managers. Because these were the managers who preferred Congolese engineers who had received training in South Africa, the managers' departure in the early 2010s contributed to the collapse of the returned migrants' supply networks among Congolese engineers (Mongala, Engineer, Wangphone, Personal Interview, Kinshasa, January 2017).

Interestingly, though, changes in ownership and deteriorating financial situation did not affect other telecommunications companies such as Mzansiphone. Mr. Tuana, Senior Human Resources Executive at Mzansiphone, explained how at his company engineers worked in the Technical Department. This is an important department in the telecommunications sector because it is the heart of a telecommunication company. Employers in the sector generally prefer hiring trusted high-tech employees here. For that reason, when there is an ownership change, new employers tend to bring with them their own engineers, replacing those of the former owners (Mr. Tuana, Senior Human Resources Executive, Mzansiphone, Personal Interview, Kinshasa, July 2017). Mr. Tuana's view corroborates Gomes et al. (2013) view on Vodacom, which sent its Congolese technicians and engineers for retraining in South Africa when it moved into the DRC in 1999. On the other hand, employers who maintain ownership of their companies tend to retain trusted engineers and invest in upgrading their skills. Mzansiphone is a case in point, because its ownership did not change since the late 1990s (Tuana, Senior Human Resources Executive, Mzansiphone, Personal Interview, Kinshasa, July 2017). One of its engineers, Tele, indicated that he and his colleagues would not be panicked by outsourcing as long as their employers were still the same South Africans (Tele, Engineer, Mzansiphone, Personal Interview, Kinshasa, March 2017).

The outsourcing trend, however, manifested scamming and the exploitation of telecommunications workers too. The research found evidence that outsourcers deceived their employees to hold erroneous conceptions of their professional status. Thus three young women namely, Busira, Bushimay and Bangamelo worked as freelancers but informed interviewers that they were permanent employees and had signed official contracts with their companies (Busira, Freelancer, Mzansiphone, Personal Interview, Kinshasa, February 2017; Bushimay, Freelancer, Yugphone, Personal Interview, Kinshasa, January 2017; Bangamelo, Freelancer, Wangphone, Personal Interview, Kinshasa, February 2017). An explanation by Maringa, a Senior Human Resources Manager at the company, confirmed these respondents' assertions:

> We've never hired a single *freelancer*. (Maringa, Senior Human Resources Executive, Yugphone, Personal Interview, Kinshasa, June 2017)

Luenda, a Junior Human Resources Executive at Mzansiphone, confirmed that this practice was quite common with outsourcers:

> Outsourcers often deceive their employees. It happens like that in many companies. (Luenda, Junior Human Resources Executive, Mzansiphone, Personal Interview, Kinshasa, July 2017)

Labour Allocation

From the foregoing, it is clear that employees could position themselves better in the workplace if they got hired and appointed through worker networks. However, this did not translate into job security because economic, professional and organisational dynamics could weaken the networks. Importantly, though, these employees' hierarchical ranks remained the same when they got outsourced despite working conditions deteriorating.

In sum, through workers' supply networks, engineers and architects in the construction, retail and telecommunications sectors in the DRC got jobs on fixed-term contracts with good pay and enjoyed favourable working conditions. The engineers and architects especially in the construction sector created a professional enclave that helped them during hiring processes. People who did not belong to that professional enclave tended to be more exposed to unlawful hiring processes and practices. The retail sector shows different patterns. Some managers got appointed through transnational labour allocation processes while others used a mixture of strong and weak ties during the hiring processes. In the telecommunications sector, changes of ownership seriously affected the hiring of professionals. A particularity of that sector is that employees at the bottom of the wage ladder bore the burden of outsourcing.

The foregoing examination also shows the usefulness of Peck's notion of labour incorporation and allocation, which postulates that, in addition to human capital, employers'/workers' networks, labour market institutions and state actors play an important role in employee recruitment and hiring processes. As the data presented in the chapter shows, these factors played a significant role in the hiring procedures that

Congolese employees went through. In the specific context of the post-war labour market of the DRC, the recourse to the different networks varies according to an employee's social class or socio-professional category. The data we presented here shows that if an employee belonged to a powerful professional, political or family network, the more they were likely to be hired in an interesting position in their company. The findings also nuance Granovetter's dualism by demonstrating how strong ties and weak ties can be simultaneously effective during hiring processes.

Conclusion

This chapter analysed the insertion of Congolese employees into the private sector, particularly in companies owned by South African, Indian and Chinese businesses in the DRC's construction, retail and telecommunication sectors. The chapter argued that most Congolese employees faced difficulties during their integration into the labour market because of high levels of irregularity in recruitment and hiring processes. The chapter developed three key sub-arguments to flesh out this argument.

First, it used Peck's notion of labour incorporation to argue that most Congolese employees found information about job opportunities through informal rather than formal processes. Equally important, it showed that many employers did not follow the nation's legal procedures in advertising job vacancies. Drawing on Granovetter's concept of weak and strong ties, the chapter showed that job seekers generally used strong ties rather than weak ties in finding jobs. Yet, these ties were not mutually exclusive in some job search processes. The chapter highlighted the use of employers' recruitment networks and workers' supply networks as important strategies for finding interesting job positions. It also demonstrated that transnational labour allocation of managers might be one exceptional feature of hiring processes in the DRC's labour market. Importantly, the chapter shed light on the paradox around the continued use of informal means during recruitment and appointment processes despite the founding of ONEM. In theory, ONEM was supposed to regulate or even solve problems related to the professional integration of Congolese into the private sector labour market. In practice, ONEM failed to check the perpetuation or escalation of informality in labour market processes of incorporation and allocation. Nevertheless, the chapter acknowledged that ONEM was a new institution that faced serious challenges because of the DRC's political, economic, administrative and sociocultural environment. All these problems prevent ONEM from working better and had negative repercussions on job seekers. Many job seekers became victims in precarious jobs or found themselves at the mercy of ruthless outsourcers.

Second, the chapter argued that few informants could get full-time permanent jobs. It demonstrated that outsourcers had shaped the hiring processes of many part-time employees. Full-time permanent employment usually resulted from employer/worker supply networks. Worker supply networks emerged as the best way of being formally hired because they helped employers find suitable job candidates quickly and appointed them in labour positions that matched their abilities. In such circumstances, employees could negotiate better employment contracts. Worker supply networks also protected their members against labour market competition by creating professional enclaves.

Third, the chapter made it clear that the match between employees' profiles and their jobs varied according to the skills required in a given sector and/or the fact of belonging to employers/employees' networks. It showed that the more employees possessed desirable skills in companies, the more they held jobs that corresponded with their human capital. Engineers, architects and technicians were placed in job positions that corresponded with their profiles in the construction and telecommunications sectors. The chapter showed how belonging to networks very close to employers allowed certain employees to be placed in executive positions in retail companies. Nevertheless, the vast majority of Congolese employees found themselves underemployed for fear of the alternative, unemployment.

Ultimately, the chapter theoretically contributes to the literature on employees' integration into the labour market. The chapter highlighted some special ways in which some people in the DRC got integrated into the post-war labour market. It also challenged Granovetter's theory that clear boundaries exist between weak ties and strong ties in job search processes. The chapter showed that these ties can mutually interact. It additionally showed that employers can initiate and lead processes of advertising job vacancies by directly contracting needed employees in their companies. Finally, the chapter demonstrated that the establishment of national employment services of placement to improve access to jobs is not necessarily synonymous with contributing to the reconstruction of labour markets in a post-war context. As long as the services remain weak, they are useless in a post-war labour market.

References

Akanji, Babatunde. 2016. "The Bleak Nature of Hrm Practices in Organisational Management of Casual Labour Employments." *Business Excellence and Management* 6(3):5–15.

Berg, Janine. 2015. "Labour Market Institutions: The Building Blocks of Just Societies." Pp. 1–35 in *Labour Markets, Institutions and Inequality. Building Just Societies in the 21st Century*, edited by J. Berg. Cheltenahm, Northampton, Geneva: Edward Elgar and International Labour Office.

Fernández-Aráoz, Claudio. 2001. "The Challenge of Hiring Senior Executives." Pp. 182–206 in *The Emotionally Intelligent Workplace*, edited by C. Cherniss and D. Goleman. San Francisco: Jossey-Bass.

Gomes, Emanuel, Marcel Cohen, and Kamel Mellahi. 2013. "When Two African Cultures Collide: A Study of Interactions Between Managers in a Strategic Alliance Between Two African Organizations." Pp. 180–201 in *Effective People Management in Africa*. Springer.

Granovetter, Mark S. 1973. "The Strength of Weak Ties." *American Journal of Sociology* 78:1360–80.

Granovetter, Mark, and Charles Tilly. 1987. "Inequality and Labor Processes." In *Handbook of Sociology*, edited by N.J. Smelser. Thousand Oaks, CA: Sage.

Inaka, Saint José. 2020. Post-War Labour Market Reconstruction: The Case of the Democratic Republic of the Congo, PhD dissertation, Sociology, University of Pretoria.

Mathekga, M. J., and Loyiso Maciko. 2018. "The Retail Sector Growth, Expansion, Employment and Income Inequality: A Case Study of Shoprite and Pick'n Pay in South Africa and Namibia." *Journal for Studies in Humanities and Social Sciences* 7(7):37–61.

Muller-Camen, Michael, et al. 2001. "Between Home and Host Country: Multinationals and Employment Relations in Europe." *Industrial Relations Journal* 32(5):435–48.

National Agency for the Promotion for Investment. 2016. Hiring and Dismissal of Workers ANAPI, ed: ANAPI.

ONEM, Office National de I'Emploi. 2015. Rapport National sur l'Emploi. Base sur le modele adopte par la SADC. T.e.P.S. Ministere de l'Emploi, ed. kinshasa: Ministere de l'Emploi, Travail et Prevoyance Sociale.

Ordre National des Architectes. 2019. Chronogramme d'activités de l'Ordre National des Architectes (ONA) de la R.D. Congo, Vol. 2019.

Peck, Jamie. 1996. *Work-Place: The Social Regulation of Labor Markets*. New York: Guilford Press.

Pruthi, Sarika, and Mike Wright. 2017. "Social Ties, Social Capital, and Recruiting Managers in Transnational Ventures." *Journal of East-West Business* 23(2):105–39.

République Démocratique du Congo. 2002. *Loi n°015/2002 portant Code de Travail*. Kinshasa: Journal Officiel de la Republique Democratic du Congo.

République Démocratique du Congo. 2017. *Loi No. 17/001 du 08 fevrier 2017 fixant les regles applicables a la sous-Traitance dans le secteur prive*. Kinshasa: Journal Officel.

République Démocratique du Congo. 2018. *Loi No. 18/034 du 13 décembre 2018 portant création, organisation et fonctionnement de l'Ordre National des Architectes*. Kinshasa: Journal Officiel.

Stewart, Frances. 2016. *Horizontal Inequalities and Conflict: Understanding Group Violence in Multiethnic Societies*. New York: Palgrave Macmillan.

Webster, Edward. 2005. "Making a Living, Earning a Living: Work and Employment in Southern Africa." *International Political Science Review* 26(1):55–71.

PART III

Empirical and Theoretical Contributions to the Study of Labour Markets and Post-War Labour Market Reconstruction

CHAPTER 8

The Burden of Getting a Job in Reconstructed Post-war Labour Markets: Some Concluding Remarks

This book sought to understand the actions and involvement of key actors in processes of post-war labour market reconstruction in the Southern African Development Community (SADC) region. It examined the reconstruction of the labour markets and went on to highlight and draw on the specific reconstruction experience in these processes of a SADC member state, the Democratic Republic of the Congo (DRC). The DRC experience considered the remaking of the country's post-war labour market after the 1998–2002 war in that country.

The Regional Context of Post-war Labour Market Reconstruction

To better locate the case of the DRC in the Southern African regional context, the book started with an overview of post-war reconstructions in SADC member states that previously engaged in wars of liberation from colonial rule and/or apartheid and civil wars that occurred in the post-colonial era. The book analysed the historical evolutions of labour markets in Angola, Mozambique, Namibia, South Africa and Zimbabwe. It considered the fact that imperial governments and colonial settlers once occupied and dominated these countries. Colonialists subsequently and successively created slave labour markets, forced labour markets and

S. J. Inaka et al., *The Reconstruction of Post-War Labour Markets in The Southern African Development Community*,
https://doi.org/10.1007/978-3-031-37357-2_8

racialised dual labour markets. The labour discriminations that characterised these markets precipitated wars and conflicts in these countries during and/or after the colonial era.

Further to analysing processes of post-war labour market reconstructions and the synoptical view of the remaking of post-war labour market in the SADC, the book zoomed in and focused on labour market reconstruction processes in the DRC. In so doing, the authors sought to bring in empirical evidence on these processes in support of the arguments of the book. We thus put the key arguments into conversation with the empirical evidence and views of people who operate in the labour market of the DRC and underscore the empirical contributions to the scarce literature on post-war labour market reconstructions in Sociology, Economics, Politics, Law, International Relations, various areas of African Studies and other disciplines.

In relation to the DRC, we argued that its labour market went through a long period of post-war reconstruction, marked by delays and/or poor implementation of labour market policies. We drew on Peck's (1996) theory of the social regulation of labour markets and asserted that the reconstruction process was shaped by a comprehensive range of dynamic factors related to the international system, local politics, state governance and conflicting interests of the actors involved. We then presented six specific arguments that support this assertion.

The Domestic Context of a Nation-State's Post-war Labour Market Reconstruction

Continuities and Raptures in Post-war Labour Market Reconstruction

Firstly, the labour market of the DRC should be understood in relation to the country's history and the existence of both continuities with and ruptures from past realities. We gave an historical account of how dynamics involving the international system, various political regimes in the DRC and power relations within those regimes shaped the country's labour market. We showed that King Leopold II of Belgium instituted and regulated a cruel labour force regime in the country from 1885 to 1908 and that the regime was ended by the international community. This led to the implementation and regulation of a dual labour market regime in the Belgian Congo from 1908 up to independence in 1960. The international context of the Cold War, marked by ideological conflict between capitalism and communism, influenced three secessions and rebellions in the 1960s that split the DRC into four and created separate labour markets. The Congolese government led by then President Joseph Kasa-Vubu (1960–1965) and the then under General Mobutu (1965–1997), backed by capitalist allies, emerged victorious from these armed conflicts, reunified the Congo and reconstructed it as well as its labour market. President Mobutu's early political, economic and labour market reforms ameliorated the labour market until 1974, though they did not effectively address the persistent colonial anti-trade unionist sentiments. Additionally, labour discrimination practices evolved, from a colour-bar labour market during the colonial era, into a politicised and *ethnicised* labour market that was also characterised by the unlawful involvement of non-market actors. The economic and political landscape of the DRC increasingly deteriorated from 1973 up until the collapse of the Mobutu regime in 1997. President Laurent-Désiré Kabila could not really reconstruct the DRC's labour market after the 1996–1997 war, but rather maintained and aggravated a 'Mobutist labour market'. This could be explained by the fact that his regime was autocratic as well as by the 1998–2002 war. Many unfair labour market practices inherited from the colonial period and/or the early post-independence regime of Mobutu constitute historical legacies that persisted well into the period after the 1998–2002 war.

Laggard Implementation of Labour Reconstruction Policies

Secondly, the state in the DRC was very slow in implementing policies and activities meant to reconstruct the country's post-war labour market. This was besides poor implementation of policies. Related to this, we argued that the neoliberal policies for post-war rebuilding imposed on the DRC by international financial institutions and the donor community and the 2003–2006 power-sharing transition regime were not conducive to a functioning labour market. They led to the legally sanctioned politicisation of the labour market that prevented the implementation of any labour market reforms during this period. The situation did not improve in the 2007–2011 period that was characterised by imbalanced power relations within the political class. From 2011, the government proceeded, slowly and uncertainly, with labour reforms that included the introduction of the *bancarisation* policy in 2013. However, procedural flaws, inefficient administration and bureaucratic inertia made the implementation of these reforms difficult. In 2015 government representatives and international partners of the government publicly admitted that much more needed to be done. Overall, the reconstruction of the DRC's labour market has thus lagged since the end of the 1998–2002 war. This argument is supported by statistical data on the DRC's labour market. The data shows that some challenges exist in the reconstruction of the labour market in the post-war period. Besides the data, we also drew on Peck's (1996) argument that the labour market is, above all, a political construction. This explains the state's cumbersome progress in reconstructing the post-war labour market in the DRC. We actually pointed to the lengthy process of implementing a minimal wage policy to underscore and also support this argument.

Insidious Political and Socioeconomic Obstacles to Reform

Thirdly, and in relation to examining the role of labour market institutions in facilitating the reconstruction of the post-war labour market in the DRC, we asserted that a set of political factors prevented these institutions from contributing to the reconstruction of the labour market in a strong and sustained manner. This was especially so during the period of the 2003–2006 transition government. We explained that corruption, weak public administration and a neo-patrimonial state prevented labour market institutions from leading an effective recovery. We additionally highlighted the existence and actions of non-market actors as major constraints on labour market institutions.

Minimal Impact of Private Sector Institutions

Fourthly, we posited that private sector labour market institutions did not sufficiently contribute to the reconstruction of the post-war labour market in the DRC. To this effect, we gave an account of the struggle between employer organisations and unionists in the country since the colonial period. With this, we showed how employer organisations have always adopted strategies to retain the state as their ally and ensure that the labour market worked in a manner that satisfied their interests. Although they created jobs, employer organisations also aimed to exploit the workforce for business purposes.

Despite changes in labour market regimes, the exploitation of workers seems to have always persisted in the DRC. The organisation that should have and claimed to have helped workers in this situation, the *Fédération des Entreprises du Congo* or Congolese Business Federation (FEC), prevented the execution of measures that favour workers. It did so with the support of influential Congolese politicians. FEC efforts to help workers deal with their misfortunes were, thus, quite futile.

The Congolese labour market has also always been unfavourable to unionism. Unions fight against a two-headed enemy in the form of an alliance of state and employers. In the post-war period, a set of basic themes continues from the past: abuse by politicians and employers, the proliferation of bogus unions, high levels of unemployment and most of all, the lack of trust in unions by their own members. Nevertheless, unions try to contribute to the democratisation of the labour market in the post-war period, despite being limited by unconducive conditions in the country. We thus drew on Peck's (1996) notion of the labour market as a fighting arena to posit that, unionists in the private sector struggled to make a difference in the process of rebuilding the post-war labour market in the DRC.

Strength in Informality

Fifth, the entry of Congolese job seekers into the private sector labour market and particularly private sector companies owned by South Africans, Indians and Chinese and operating in the construction, retail and telecommunications sectors. We deployed Peck's (1996) notion of labour incorporation and allocation to show that most Congolese job seekers obtained information about job opportunities through informal processes. We also employed Granovetter's (1973) concept of weak and strong ties to show that Congolese job seekers had more recourse to strong ties than weak ties in the quest for or exposure to information on job opportunities.

More interestingly, our research showed the existence of mixed ties in the DRC. We observed that some single job seekers used both strong ties and weak ties to gather information on job vacancies. To us, this meant no solid distinction existed between strong and weak ties and that ties can either overlap or not, during a job seekers' labour incorporation process. This had implications for information gathering regarding job search processes and possible consequences for post-war labour market reconstruction.

Our investigation revealed that many employers did not comply with the law either, as they did not advertise job vacancies through the National Employment Office or Office National de l'Emploi, (ONEM), whose official mandate is to link job seekers with employers. Thus, we uncovered an inconsistency between the existence of ONEM and the persistence of informal hiring processes in the post-war labour market of the DRC. We demonstrated that ONEM's weaknesses and failure to perform its role gave room for the continuation and intensification of these informal processes.

Job Mismatches and Underemployment

Lastly, our investigation showed that the companies that we focused on rarely match jobs with workers. Workers had to accept being underemployed to escape unemployment. Since 2011, full-time permanent jobs became scarce in the retail and telecommunications sectors in the DRC. This was an outcome of outsourcing and changes of ownership in some companies. Most Congolese job seekers, therefore, had enormous challenges integrating into the country's post-war labour market.

Empirical Contribution to Post-war Labour Market Reconstruction Research

A further and by now clear contribution this book also makes relates to studies on post-war labour market reconstruction by focusing on the SADC region. Our study provides evidence that labour issues triggered wars in some SADC's countries. Regarding the DRC, for example, we considered two intense periods of war: 1960–1964 and 1998–2002, respectively, and paid particular attention to the similarities of the two wars. Both led to a fragmentation of the labour market. In both, international dynamics (the Cold War in the 1960s and neoliberalism and globalisation in the new millennium) strongly contributed to the post-war recovery.

In the book, we also established that the history of the DRC's market shows continuity and discontinuity in the legacy of the past in post-war labour market reconstruction. Thus, some features of the current labour market reproduce the forced labour regime under King Leopold

II (1885–1908), the dual labour market of the Belgian Congo (1908–1960), the fragmented labour market during the Cold War (1960–1965), the turbulent labour market during the Mobutu's reign (1965–1997) and Kabila's, and the second fragmented labour market during the war of 1998–2002.

Additionally the book makes a contribution that relates to the impacts of political power sharing and labour market. We demonstrated how the legalised politicisation of the labour market, through the Pretoria Agreement and the new constitution of 2003, had adverse repercussions. We particularly highlighted the multiplication of chains of command, the blockages of reforms for the public and private sectors and the conflicts between existing labour laws and the constitution of the transition (2003–2006). Our study also focused on other repercussions, such as the perpetuation and aggravation of unfair practices inherited from the past. Here, the study went beyond the extant literature by highlighting the legal aspect of the politicisation of the labour market. It shed light on the way in which legal politicisation allowed politicians to act unfairly and justify the legality of their unfair actions in the labour market. Although politicisation of the labour market is often decried as illegitimate or even illegal, the study in this book has brought into the literature a discussion of what would constitute the legitimate and legal aspect of politicisation.

The book has presented evidence of how the activities of non-market actors harmed the rebuilding of the labour market and showed that from Mobutu's Zaïre to the 2020s, these actors exerted destabilising influences on the daily activities of the labour market. In this way, this study has broken new ground in exposing how the entry of non-market actors into labour market institutions disrupted the functioning of the market, and consequently its post-war reconstruction.

Another important contribution of the study presented in this book is its analysis of the contribution of the private sector to the post-war reconstruction of the labour market, in particular the identification of the struggle between employer organisations and unions. The analysis of this struggle offers the existing literature a clear and precise picture of the almost unlimited capacity of unions and employer organisations to disrupt labour market. Their unfair conflict and frantic pursuit of self-interest exposed workers to all sorts of misdeeds in the labour market.

A final empirical contribution is the fact that workers' networks have implicitly substituted employment structures. This substitution is due to the weakness of the mandated institutions to play this role. Our study in the book thus adds to the literature on the roles of worker networks in hiring processes and the important aspect of how the activities of these worker networks confirm the weakness or lack of institutional structures. To the best of our knowledge, the informal replacement of labour market institutions by workers' networks is still non-existent in the literature on post-war labour markets.

Theoretical Contribution to Post-war Labour Market Reconstruction Research

From a theoretical perspective, it is important to underline that the visible links between the past and the present in the labour market of the DRC and the surrounding SADC region are consistent with Peck's argument that labour markets are historically constructed and embedded. However, our study went beyond Peck's position by showing that the DRC and SADC region's labour market was historically constructed before wars, wrecked during wars and reconstructed after wars. This leads us to posit that in abstract terms, labour markets are not only socially, politically and historically constructed, but also destructed and reconstructed in particular ways. That is, labour markets are in perpetual movement of structuration and deterioration, or explosion and restructuration. They can be made before wars, unmade during wars and remade after wars, and this is for a succession of wars.

Whereas our book has characterised the changes in post-independence labour markets constructed after the wars of liberation in the SADC region as similar, nuanced differences might exist within countries that subsequently regressed into post-independence civil war in the region. Future comparative research on the latter group of countries and their experiences as well as those that did not experience post-independence civil war in the SADC region would enrich the understanding of post-war labour market reconstruction. Overall, it is important to emphasise that this book has shown the need for further studies of post-war labour market reconstruction. Its modest contributions to the limited literature on the subject cannot be complete. A window remains for further studies to deepen what the book may only have begun to sketch.

References

Granovetter, Mark S. 1973. "The Strength of Weak Ties." *American journal of sociology*, 78:1360–80.

Peck, Jamie. 1996. *Work-Place: The Social Regulation of Labor Markets*. New York: Guilford Press.

Index

S. J. Inaka et al., *The Reconstruction of Post-War Labour Markets in The Southern African Development Community*,
https://doi.org/10.1007/978-3-031-37357-2

I

J

K

L

FSC
www.fsc.org
MIX
Papier | Fördert
gute Waldnutzung
FSC® C083411

Kolibri 360 GmbH
Ferdinand-Jühlke-Straße 7
99095 Erfurt, Deutschland
produktsicherheit@kolibri360.de